Animal Rationality

Investigating Medieval Philosophy

VOLUME 12

The titles published in this series are listed at *brill.com/imp*

Animal Rationality

Later Medieval Theories 1250–1350

By

Anselm Oelze

BRILL

LEIDEN | BOSTON

The Library of Congress Cataloging-in-Publication Data is available online at http://catalog.loc.gov
LC record available at http://lccn.loc.gov/2017061201

Typeface for the Latin, Greek, and Cyrillic scripts: "Brill". See and download: brill.com/brill-typeface.

ISSN 1879-9787
ISBN 978-90-04-36362-5 (hardback)
ISBN 978-90-04-36377-9 (e-book)

To all animals

Contents

Preface

This book is about animals. About animals like you and me but also – and mainly – about the members of other animal species such as dogs, cats, apes, or ants. A topic like this may not seem very original these days, because interest in animals has grown immensely over the past few decades. It has grown to the extent that some even say that we are currently observing an 'animal turn' across various disciplines. Regardless of whether or not this is true, there is undoubtedly much research being done on animals, their mutual relationships, common history, differences, and commonalities at the moment and this interest is shared by people working in different fields, including philosophers. The philosophers' main interest in this context lies, as many people think, in animal *ethics*, that is, the question of the moral status of humans in comparison to other animals. The answers given to this question surely matter, as they affect the way in which we interact with and act on other animals. However, there is also a second field that is philosophically relevant and this is the field of animal *cognition*. One might even argue that this field is more fundamental than animal ethics since arguments regarding the moral status of animals often rely on what we think about their cognitive capacities. Therefore, it is hardly surprising that the topic of animal cognition has a long philosophical history.

In recent years, several studies have shed light on this history. Some periods, however, have received considerably more attention than others. While the views of ancient, early modern, and contemporary philosophers have been studied a great deal, the views of medieval thinkers have yet to attract as much attention as they might. A very general reason for this lack of attention is that, despite all efforts, medieval philosophy still leads a relatively shadowy existence within contemporary (history of) philosophy. Yet, there is also a more specific reason, namely, the widespread opinion that there is not much to say about animal cognition in the Middle Ages. Medieval philosophers, it is often said, granted various sensory capacities to nonhuman animals, such as sight, hearing, imagination, and memory. Still, they denied them anything related to the possession of a rational soul, such as concept formation, judging, reasoning, or prudence. Consequently, nonhuman animals are *non-rational* animals and their cognition is nothing but a deficient version of human rational cognition.

Although this view contains more than a grain of truth, it fails to capture the depth and diversity of the medieval discussion. As this study shows, medieval philosophers did not stop at the ascription of sensory cognition to other animals. Rather, they wondered whether one can coherently account for many complex behaviours in nonhuman species without granting them complex cognitive processes. In particular, the question was whether those processes that were taken to be rational processes always require rational faculties or whether they can also exist without a rational soul. This question might, of course, seem to be heretical in at least two regards. First, it seems to contradict the basic idea of Aristotelian faculty psychology, according to which every psychological process requires a particular type of soul or faculty of the soul. Rationality without reason thus seems to be impossible. Second, it seems to undermine the Aristotelian definition of the boundary between humans and nonhuman animals because, in Aristotle's view, rationality is what sets humans apart from all other animals. Yet, despite these dangers for Aristotelian psychology and anthropology (which, incidentally, went hand in hand with the Christian idea of human nature), medieval philosophers did not refrain from asking these questions and the answers they gave were quite different. It is these answers that shall be called *theories of animal rationality* and they are the subject of this book.

It is important to note, however, that this book does not cover the whole range of medieval theories from the fifth to the fifteenth centuries. Rather, it focuses on the theories of Latin authors from roughly 1250 to 1350. This period is a particularly rich and interesting period because the transmission of Aristotle's writings had been largely completed by that time. Aristotle's treatises on psychology and zoology as well as his works in other fields, such as metaphysics or ethics, gave a major boost to theorising in the Latin West. In a sense, this was a late-medieval precursor of the modern animal turn because animals became the subject of various debates not only in zoology but also in psychology and philosophy. Still, one must not forget that there are many theories that were developed before and after that period and outside the Latin cultural area, especially in the Islamic world during the eleventh and twelfth centuries. Many Islamic thinkers paved the way for the Latin discussion. Yet in the present book their theories will be considered only insofar as they had a notable influence on the ideas of Latin thinkers. This is not to say, of course, that they are less relevant or less interesting but they go beyond the scope of this book and definitely deserve a study in their own right.

The present study is a revised version of my doctoral thesis, which I defended in January 2017 at the Humboldt University of Berlin. Its writing would not have been possible without the help and support of several people and institutions. First of all, I would like to thank my supervisors Dominik Perler and Bernd Roling. Their articles on the subject were among those publications that mainly inspired me to begin working in this field. The supervision they gave me from the very first moment was brilliant and this work has profited a lot from their advice and comments. Dominik Perler's colloquium at the Humboldt University was a stimulating place of intellectual exchange and the colloquium's participants who commented on various drafts of this book helped me to improve it a great deal. Many important suggestions I also owe to audiences at Berlin, Cambridge, Helsinki, Jyväskylä, Köln, Leipzig, Leysin, Münster, Oxford, Porto Alegre, and Uppsala where I had the chance to discuss different aspects of my work over the past four years. I am particularly grateful to the following people who provided valuable materials and comments or inspired me in one way or another: Henryk Anzulewicz, Mark Barker, Manuel Bohn, Alexander Brungs, Emanuele Coccia, Tobias Davids, Silvia Donati, Thérèse-Anne Druart, Maarten J.F.M. Hoenen, Martin Klein, Theodor W. Köhler, Jörn Müller, Martin Pickavé, Paolo Rubini, Juhana Toivanen, Cecilia Trifogli, and Markus Wild. Special thanks go to Christoph Sander since he never tired of discussing my project. His astute remarks definitely saved me from making several mistakes as did the careful eyes of Bradley Burroughs and Chad Jorgenson who proofread the entire manuscript. For any remaining flaws I naturally take full responsibility.

Financial support for this project came, for the most part, from the German Academic Scholarship Foundation. The grant I received enabled me to fully concentrate on the dissertation which is a privilege one cannot overestimate. In addition, I received a generous bursary from Dominik Perler's Leibniz Prize project "Transformationen des Geistes" which gave me an easy start. For the final stage, José Filipe Pereira da Silva invited me to become a member of his ERC-funded project "Rationality in Perception: Transformations of Mind and Cognition 1250–1550" at the University of Helsinki (ERC Grant Agreement No. 637747). They thus put me in the position to begin and finish my work without financial concerns and for this I am more than grateful. I am also grateful to John Marenbon who accepted the book for publication in this series and to Marcella Mulder from Brill who was a very thoughtful and reliable editor.

Above all, I thank Minka and Nora. Minka, because she used to patiently sit on my desk when, time and again, I made her wait for lunch or dinner since I was too busy thinking over whether beings like her are rational animals. And Nora, because every day she makes me see that life is about more than rationality.

Anselm Oelze
Leipzig, November 2017

Introduction

According to a long-standing philosophical tradition, we humans are rational animals. This means that we are endowed with certain cognitive powers, namely, intellect and reason, that enable us to engage in various cognitive operations, such as concept formation, judging, or reasoning. It is these operations that shape the way in which we perceive and interact with the world. We conceptualise the brown furry thing we see as a dog, we judge that the dog is hungry when it desperately stares at the feeding bowl, and we reason that we should feed it if we want its hunger to disappear. To some extent, these cognitive operations even put us in a position to build our own worlds like, for instance, the world of logic or the world of science. In these worlds, dogs are not simply our pets or companions but they become the objects of our research. We might study their biology, physiology, and psychology and perhaps find out that dogs are just as smart and intelligent beings as we are. This finding would definitely change the way in which we treat dogs. We would begin to give them all the rights we grant to human beings and dog lead producers might finally become redundant.

As we all know, we have not yet reached this point and maybe never will. Still, many people would say that we are much closer to reaching it than our ancestors ever were. Since the advent of modern biology, psychology, ethology, and similar disciplines we have left the above-mentioned tradition behind. In its original form, or so the usual story goes, this tradition is rooted in the thought of Aristotle because it was Aristotle who claimed that only human animals possess rational souls or souls with rational faculties like intellect and reason. The denial of intellectual faculties to other animals was fateful, according to several scholars because it "produced a crisis in psychology, in ethics, and in religion," as Richard Sorabji famously put it.[1] It produced a crisis in the sense that nonhuman animals were henceforth trapped in the sphere of arationality, that is, in a sphere significantly inferior to the sphere in which our own species resides. It took more than two thousand years for this crisis to come to an end. Of course, we still put leads on dogs, keep chimpanzees in zoos, and eat cattle. Nonetheless, we have begun to understand that the Aristotelian definition of the boundary between human and nonhuman animals needs to be revised. And once this has happened, the crisis will be over.

Now one might object that there never was such a crisis in the history of Western philosophy. For not only was Aristotle one of the most important and

1 Sorabji (1993a), 1. See also Sorabji (1993b), 7f., and Steiner (2008), 17f.

 | DOI 10.1163/9789004363779_002

impressive philosophers of all times, but he also was one of the most prolific writers in what can count as the precursors of modern biology, psychology, and ethology. His treatises on the physiology and psychology of animals, both human and nonhuman, consist of several thousand pages and so he seems to be above suspicion in this regard since his picture of the animal realm is far from being one-dimensional. Be that as it may, one can hardly deny that there are many passages in Aristotle's writings in which he clearly rejects the attribution of intellectual powers to nonhuman animals.[2] It would be naïve to think that these passages did not have an influence on those thinkers who came after him. The crucial question, however, is: what did this influence amount to? Was Aristotle really the beginning of a one-way road? Did his denial of reason to nonhuman animals lead everyone after him into the same direction? Was everyone hit by the crisis?

If one looks at the history of the animal/human boundary, one easily sees that there certainly was more than just one definition of this boundary, even following Aristotle. Ironically, it was Richard Sorabji himself who, in his seminal study of the history of the Western debate, has shown that already in Antiquity there was a whole spectrum of positions concerning the cognitive capacities of animals. Cats, bees, apes, or elephants were not taken to be dumb and unintelligent by everyone. There were also many who pointed to the astonishing abilities they have, many of which they share with us.[3] However, this spectrum seems to have disappeared by the beginning of the medieval period at the latest because, according to Sorabji, the (Western) Middle Ages were dominated by a "Christian tradition which selected just one side from a much more wide-ranging Greek debate."[4] Thus, even if Aristotle's views had produced less severe a crisis than originally thought, the beginning of the medieval period marked the beginning of a "dark age" for nonhuman animals.[5] And if we still have to deal with the consequences of that crisis some say it is because of "the gulf that millennia of Judeo-Christian indoctrination have dug between us and other animals."[6]

2 See, for instance, *De anima* II.3, 414b18-19 and 415a7-8; *De partibus animalium* 641b7f.; *Politica* 1332b3-5. For a complete list of the *loci classici* see Sorabji (1993b), 12, n. 30.

3 Sorabji (1993b). Besides Sorabji's study there are also valuable works on ancient views of animal cognition by Dierauer (1977); Cassin & Labarrière (1997); Steiner (2005), esp. 38–111; Gilhus (2006); Osborne (2007); Newmyer (2011); Harden (2013); Li Causi & Pomelli (2015).

4 Sorabji (1993b), 8.

5 Preece (2002), 62–90.

6 Singer (2006), 158.

Claims like these certainly gain some plausibility from the fact that medieval Christian thinkers amalgamated the Aristotelian definition of the animal/human boundary with two particular biblical doctrines found in Gen 1:26-28. The first of these doctrines is the idea of humans being created 'in the image of God'. By and large, this was interpreted to mean that humans are the only earthly creatures with a rational and immortal soul. Because of this rational soul they are cognitively superior to other terrestrial creatures. Closely linked to this cognitive superiority is humanity's moral predominance. The doctrine of the '*imago-Dei*' was thus accompanied by the idea of what is called the '*dominium terrae*', humans' dominion over the earth, including all living beings.[7] Because of this exceptional position of humans, the chasm between us and other animals was certainly deepened. As far as the discussion of animal cognition was concerned, "[a]ny real study of animal psychology would have to wait until after the end of the Middle Ages," as Joyce E. Salisbury put it.[8] But is this actually true? Is it true that the Middle Ages were the height of the crisis diagnosed by Sorabji?

On the one hand, this picture is not entirely incorrect. For most, if not all, medieval authors subscribed to the premise that only human animals are *rational animals* in the sense that only they possess rational souls or the faculties of intellect and reason. Furthermore, medieval thinkers did not devote the same amount of attention to the discussion of nonhuman animals' cognitive capacities as to other fields or disciplines from logic and metaphysics to Christology. But, on the other hand, it would be equally wrong to think that they did not care about this at all. In fact, they had a keen interest in finding out what separates humans from other animals, and they did not hesitate to ask whether the cognitive capacities we have are so much different from the capacities of other animals. They wondered what kind of cognitive processes underlie the behaviour of dogs, cats, bees, sheep, monkeys, ants, and many others, and they asked how such processes compare to those underlying our own behaviour. Are they entirely distinct or are they similar? Or are they even equal, perhaps? And how can we find this out if all we can do is observe other animals' behaviour?

As the present study aims to show, not one but many answers were given to these questions. The spectrum of medieval views on the cognitive capacities of nonhuman animals was much more varied and disparate than one might think from hearing the story of the crisis. And so the denial of intellect and reason was not a one-way road. It led in different directions and, in particular,

7 For a brief overview of these doctrines see Baranzke (2002), 82–88. For a critical discussion of the history of their reception see Cohen (1992).

8 Salisbury (2011), 5.

to the development of what can be called *theories of animal rationality*. This might seem like a plain contradiction in terms. If there really was so much support for the view that only humans are endowed with intellect and reason, it seems odd, if not wrong, to say that medieval thinkers developed something like theories of (nonhuman) animal rationality. But the point is that there was not much of a debate about whether nonhuman animals have *cognition*. It was relatively uncontroversial that almost all animals share a certain number of external and internal senses, such as sight and hearing or imagination and memory. What was at issue was rather the kind of cognition they have and whether their cognition is entirely *non*-rational. Although one might think that this question was taboo in the Middle Ages, the present book shows that it was not. Despite all of the differences between the present and the medieval debate on animal rationality, there are also many astonishing parallels. This is something one is likely to miss when hearing the story about the fateful crisis produced by Aristotle.

CHAPTER 1

What are and Why Study Later Medieval Theories of Animal Rationality?

Contemporary theories of animal rationality concern a relatively wide range of issues. They range from inferential and linguistic capacities to social and meta-cognition.[1] What is common to most of them is that they are interested in *process* rather than *behavioural rationality*.[2] Generally speaking, these are different ways of telling whether or not a being is rational. When we focus solely on its behaviour and say that it is rational, since, for instance, it chooses the right means to achieve a certain end (and so 'maximises utility', as some people put it) we ascribe behavioural rationality to that being. According to this concept of rationality, almost any kind of being could count as rational in one way or another. Not only people can be rational but also institutions, machines, or genes as long as they *behave* in a way that qualifies as rational.[3] What matters is simply whether or not a certain behaviour can be described as rational. If this is the case, then the being exhibiting this behaviour can be called rational.

However, this type of rationality is much more interesting and relevant for economists, for example, than it is for psychologists. The latter are interested rather in the question whether the behaviour we observe is based on rational *processes*. To put it slightly differently, they wonder *how* a certain kind of behaviour is brought about. For instance, they examine whether a certain kind of behaviour that can be described as rational from the outside is actually based on a rational cognitive process such as reasoning. To be clear, this does not mean that psychologists or anyone else interested in process rationality work with a better or more appropriate concept of rationality than economists and people interested in behavioural rationality. They simply look at rationality from a different angle, and when they identify rationality they do this at a different level.

In the Middle Ages, one can also find this distinction between behavioural and process rationality – not literally, of course, but systematically. A passage that neatly illustrates this is found in Thomas Aquinas' *Summa theologiae*.

1 The most comprehensive and insightful volume on these theories is Hurley & Nudds (2006a). See also Perler & Wild (2005) and Lurz (2009).

2 See Hurley & Nudds (2006b), 5f.

3 See Kacelnik (2006), 90–93.

 DOI 10.1163/9789004363779_003

When we look at how things behave, Aquinas says, we might get the impression that not only humans but also artefacts, such as arrows or clocks, are rational. An arrow, for instance, seems to behave rationally by moving directly towards its target.[4] Taking up the above-mentioned terminology one could say that the arrow shows behavioural rationality. It chooses the right means for achieving a certain end. However, Aquinas stresses that this is somehow beside the point, because what really matters when we try to identify whether or not something is rational are the following two criteria. First, we need to ask whether a certain behaviour is based on a rational process. Thus, like contemporary psychologists Aquinas looks for *process rationality*. In this sense, an arrow is not a rational being because its moving towards the target is not based on a rational process.[5] The arrow does not reason that it needs to fly in this or that direction in order to hit the target. Rather, it is the shooter that engages in this sort of cognitive operation. Yet, one might ask why Aquinas (and other medieval authors) think the shooter is capable of doing this. How is it that the shooter can engage in a process such as reasoning, while the arrow cannot?

The answer, in short, is that the shooter is endowed with certain rational *faculties*, namely, intellect and reason. This is the second thing one needs to take into account when talking about rationality, according to Aquinas. What lies behind this point is the Aristotelian theory of the soul. According to this theory, one can only engage in a certain process if one is endowed with a particular power or faculty. For things like digestion or growth one needs vegetative powers, sensing requires sensory powers, and rational cognition demands rational powers.[6] So besides behavioural and process rationality there seems to be a third type of rationality for medieval Aristotelians and this could be called *faculty rationality*. Very likely, it is because of this concept of rationality that Aristotle and his medieval followers have such a bad reputation these days when it comes to the question of animal rationality because it establishes

4 Thomas Aquinas, *Summa theologiae* I–II, q. 13, a. 2, ad 3, ed. Leonina VI (1891), 99: "[…] in omnibus quae moventur a ratione, apparet ordo rationis moventis, licet ipsa rationem non habeant: sic enim sagitta directe tendit ad signum ex motione sagittantis, ac si ipsa rationem haberet dirigentem. Et idem apparet in motibus horologiorum, et omnium ingeniorum humanorum, quae arte fiunt." For a more detailed analysis of this passage see Chapter 19.

5 It is not even minimally rational in the sense in which Dretske (2006), 107, defines it because, in his view, "[m]inimal rationality requires that thought be involved in the process by means of which the behaviour is produced." The arrow's movement is, however, not based on any kind of thought or mental representation.

6 On Aristotelian faculty psychology see the articles in Perler (2015a). On its medieval Latin reception in particular see King (2008) and Perler (2015b).

what is usually called a very strong 'anthropological difference'. This means that it clearly defines "features of homo sapiens that (a) set us apart 'categorically' or 'essentially' from all other animals; (b) are fundamental, in that (all) other relevant differences derive from them; (c) are important, notably to our self-image, for instance because they assure us of a higher spiritual or moral status than animals," as Hans-Johann Glock summarises the key features of an anthropological difference.[7]

For Aristotelians in general and later medieval Aristotelians in particular, reason does exactly this kind of job: (a) It establishes a *metaphysical difference* between humans and other animals because only humans are endowed with immaterial rational souls. (b) It establishes a *cognitive difference* because only these souls are endowed with what could be called the 'triad of intellectual operations', including concept formation, judging, and reasoning (see Chapter 6). (c) It establishes a *moral difference* because only rational beings can act for reasons, hence be morally responsible for what they do. Moreover, only such beings can be redeemed from sin, not at least because only their immaterial souls survive the death of the body. In sum, medieval Aristotelians defend a version of rationalism in the sense that for them reason is what marks the decisive difference between humans and other animals – metaphysically, cognitively, and morally.[8] Therefore, it seems that there exists nothing like a medieval version of theories of animal rationality because rationality is exactly what sets humans apart from nonhuman animals.

As mentioned before, this conclusion fails to capture the depth and diversity of the medieval discussion. Most importantly, it ignores the fact that later medieval thinkers were well aware of what can be called the 'grey areas' of Aristotelian faculty psychology (see Chapter 7).[9] These areas become obvious in those cases in which an animal that is non-rational by definition shows a certain behaviour that, when seen in a rational being, is explained by its rationality. A classical example that illustrates such a case is the example of 'Chrysippus' dog' (it is this example, by the way, that Aquinas discusses in the

7 Glock (2012), 109. See also Glock (2016b), 17.

8 On this concept of rationalism see Wild (2006), 5f. Note, however, that defenders of this kind of rationalism usually put much emphasis on the capacity of language. In the medieval period, this relation exists as well but is less prominent in the discussion of animal rationality, as we shall see.

9 I adopt this term from Jones (2013), 63, who, in his concise chapter on man and nature in the Middle Ages, remarks that "[i]f rationality could be used as a defining marker of humanity there remained a number of grey areas." In particular, he has in mind the (apparently) rational behaviours of human infants and nonhuman animals.

passage mentioned above).[10] A dog chases a deer but suddenly loses sight of it. It comes to a fork with three paths and sniffs around the first and the second path. It then takes the third path but without having sniffed around it. The question is how to explain the dog's behaviour. In the case of a human being, the explanation would go like this: the decision to take the third path arises from a rational operation such as 'The deer must have taken path A, B, or C. If there is no trace of it at path A and B, then it must have taken C'. To put it in logical terms, the human being employs a disjunctive syllogism and this is what causes its behaviour. But what about the behaviour of the dog? How can one explain that the dog shows the same behaviour as a human being in this situation?

One possible answer is that the dog's behaviour derives from the same kind of cognitive process as the corresponding human behaviour. If this human behaviour is explained by a process such as reasoning the dog's behaviour should be explained by reasoning as well. The argument behind this answer is commonly known as the 'argument from analogy'.[11] In short, the argument goes like this: If we observe the same effects, that is, the same behaviours in two different animals we must assume that these effects are brought about by the same causes, that is, the same cognitive processes. So when we see a dog solve a certain problem that humans normally solve by reasoning we must assume that the dog solves it by reasoning, too. Two prominent defenders of this argument in the history of Western philosophy are Michel de Montaigne and David Hume. Both agreed that from like (behavioural) effects one should infer like (cognitive) causes, even across species.[12] Therefore, it is hardly surprising that they made a strong case for the rationality of nonhuman animals.

This kind of argument, however, seems to be unavailable to medieval thinkers because it is incompatible with the Aristotelian model of the soul. To argue that both humans and dogs solve problems by reasoning runs counter to the idea that a process such as reasoning requires a rational faculty, the possession of which is exactly what separates humans from dogs. Surprisingly, though, later medieval thinkers show much more flexibility than one is usually inclined to expect. Instead of discarding the argument from analogy right away, they employed different explanatory strategies to deal with the problem of grey areas

---

10 On the origin, history, and reception of this example, see Floridi (1997). A recent philosophical analysis and discussion is provided by Rescorla (2009). On Aquinas's discussion see Chapter 19.

11 See Wild (2013), 84–89.

12 Although both Montaigne and Hume defend the argument from analogy the ways in which they spell it out differ slightly; see Wild (2006), 248.

in Aristotelian psychology (see Chapter 32). Some (e.g. Thomas Aquinas and John Duns Scotus) chose what is called the strategy of 'differentialism' in modern animal philosophy. That is to say, they tried to maintain the anthropological difference that is established by the Aristotelian model of the soul. In those cases where this model came under attack, for instance, in virtue of the similarities in the behaviours of human and nonhuman animals, they searched for explanations that avoid the argument from analogy and trace similar behaviours back to different cognitive causes. Even though the behaviour of Chrysippus' dog looks rational from the outside and so might qualify as behaviourally rational it is not rational from the inside, that is, neither with regard to the faculties nor with regard to the processes that bring it about.

Still, this kind of explanation was not accepted by everyone. As will become clear, there are several cases in which later medieval authors chose the strategy that is now generally called the strategy of 'assimilationism'. Instead of maintaining, if not even deepening the anthropological difference, they somehow assimilated humans and other animals. To be clear, this does not mean that they abandoned the basic principles of Aristotelian faculty psychology. Most importantly, they did not reject the denial of rational faculties to nonhuman animals. Nevertheless, some of them considered the option of what might be called 'rationality without reason' (see Chapter 33). This means that they in some sense dissolved the link between process and faculty rationality. They held that at least some nonhuman animal species can engage in rational processes, such as reasoning, although they lack rational faculties. How this is actually possible was the matter of some debate, but there were different solutions available. Albertus Magnus, for instance, argued that certain inner senses of some highly-developed nonhuman animals can engage in simple forms of reasoning (see Chapter 21).[13] He thus upgraded the powers of the sensory soul. Nevertheless, he tried to maintain the metaphysical difference between nonhuman animals' material sensory souls and the immaterial rational souls of humans. John Buridan went even further by claiming that this metaphysical difference does not exist (see Chapter 22). In his opinion (which was inspired by Alexander of Aphrodisias), all souls are material forms. He did not deny that human cognition is superior to nonhuman animal cognition in various respects but cognitive performance depends on material complexity. Hence, those animals whose souls are (almost) as complex as those of humans can engage in similar cognitive operations.[14]

---

13 On this see, for instance, Roling (2011), esp. 229–233; Köhler (2014), esp. 388; Tellkamp (2016).

14 See Pluta (1996), 92–97, and (2015), esp. 281–286.

What this variety of explanations shows is that later medieval authors did not necessarily select "just one side from a much more wide-ranging Greek debate," as claimed by scholars like Sorabji.[15] Indeed, the medieval Latin discussion differs from the ancient Greek debate in various respects. One of the main reasons for the differences is that many of the materials that covered different parts of the spectrum of the Greek discussion were not available to Latin medieval authors. Still, there was a vivid debate about whether or not nonhuman animals can be rational in the Middle Ages, too, and from this debate arose various theories of animal rationality.

The landscape of those theories has not yet been mapped and it is the aim of this book to provide such a map. It will do this by combining primary sources from roughly 1250 to 1350 (see Chapter 2 below) and secondary literature, both systematical and historical. As far as the latter is concerned, several studies that have been published in current scholarship provide an excellent starting point. They include works on medieval anthropology and animal cognition covering a wider range of views[16] as well as studies on the positions of particular authors, from Albertus Magnus,[17] Roger Bacon,[18] and Thomas Aquinas[19] to Peter of John Olivi[20] and John Buridan.[21] This list might actually create the impression that the aforementioned gap in the scholarship does not exist. And, indeed, the growing cross-disciplinary interest in animals has generated increasing interest not only in ancient[22] and early modern positions[23] but also

15 See p. 2 n4.

16 Among the most substantial and comprehensive studies on medieval anthropology are Köhler (2008) and (2014). Köhler (2000) also forms part of this impressive project. On animal cognition in the late ancient and early medieval period see Nitschke (1967); Clark (1998); Matthews (1999); Brittain (2002). Later centuries are covered by Sobol (1993); Perler (2006), (2012b); Roling (2011), (2013a), (2013b). General overviews of animals in medieval philosophy provide De Leemans & Klemm (2007); Jones (2013), esp. 62–72; Toivanen (forthcoming).

17 See, for instance, Balss (1954); Hünemörder (1980); Resnick & Kitchell (1996); Guldentops (1999) and (2000); Köhler (2001); Anzulewicz (2009); Tellkamp (2013) and (2016).

18 Wood (2007) and Hackett (2013).

19 The most recent monograph on Aquinas is Davids (2017) who substantially revises Barad (1995). Articles on Aquinas include Deeley (1971); Hünemörder (1988); Manzanedo (1989) and (1990); Yamamoto (1998); Loughlin (2001); Juanola (2015).

20 Toivanen (2011) and (2013a).

21 Zupko (2008); Pluta (2015).

22 See p. 2 n3.

23 The prime example is undoubtedly Wild (2006) who covers Montaigne, Descartes, and Hume. Studies or volumes with a broader range include Fudge (2006); Muratori (2013); Muratori & Dohm (2013); Roling (2014); Meo-Ehlert (2015); Buchenau & Lo Presti (2017).

in medieval ones. Nonetheless, there is no single monographic study which provides a historical reconstruction as well as a systematic discussion of medieval theories with a particular focus on rationality. Such a study can aid those interested in the history of philosophy as well as provide interesting insights for anyone engaged in the present debate, because by looking at the theories from the medieval period we might see more clearly what characterises the theories of our own age.

CHAPTER 2

How to Study Later Medieval Theories?

The present book focuses on *later* medieval theories, more precisely, the theories of authors who lived and worked at some point between roughly 1250 and 1350. The main reason for concentrating on this period is that it is characterised by a comparatively large amount of discussion about animal cognition in general and animal rationality in particular. One of the pivotal driving forces behind this was the transmission Aristotle's writings in the Latin West. Although it had begun much earlier, the translation of decisive parts of the *Corpus Aristotelicum* had not been completed before the second half of the thirteenth century. This body of work contains two major collections of books that are relevant in the present context (see Chapter 4). First, there is the so-called 'Books on Natural Philosophy' (*libri naturales*). Among these are the *Physics* as well as the highly influential treatise *On the soul* (*De anima*). Moreover, they include a number of 'Little Treatises on Natural Philosophy', especially psychology and physiology, known as *Parva naturalia*. The second relevant collection, the 'Books on Animals' (*De animalibus libri*), unites at least three zoological treatises: the 'History of Animals' (*Historia animalium*), 'On the Parts of Animals' (*De partibus animalium*), and 'On the Generation of Animals' (*De generatione animalium*). Together these collections fuelled the discussion on animals' cognitive capacities because commenting on Aristotle's works belonged to the principal tasks of (late) medieval academics.

There was, of course, also a discussion on animal rationality before this commentary tradition began to flourish in the Latin West. One of the most interesting figures in this respect is Adelard of Bath.[1] In his *Questiones naturales*, written during the first half of the twelfth century, he develops an interesting and quite original account of the nature and capacities of animal souls.[2] Undoubtedly, his interest in this topic arose from the far-reaching "discovery of nature" in the twelfth century (see Chapter 4). However, unlike thirteenth- and fourteenth-century theories, Adelard's account is not connected to the Aristotelian model of the soul. Just as little is it influenced by the so-called theory

1 On Adelard as natural philosopher see Burnett (1987) and Grant (2007), 117–122. On his account of animal rationality see Roling (2011), 228f.

2 See Adelard of Bath, *Questiones naturales*, qq. 13–14, ed. Burnett (1998), 110–118. Another original and perhaps even slightly earlier text that has recently been edited is Ralph of Battle, *De nesciente et sciente*, lib. I, c. 4, §§37–76, eds. Niskanen & Goebel (2015), 258–279.

 | DOI 10.1163/9789004363779_004

of inner senses. In its most influential form, this theory was developed before the thirteenth- and fourteenth centuries and outside the Latin West, especially by Avicenna. It is more than just the passive reception of Aristotle's psychology and very likely became as influential as it did in later centuries because it provided solutions to some of the problems produced by Aristotle's model of the soul.[3] Therefore, the influence of (earlier) medieval Arabic psychology and zoology cannot be ignored in a study of later medieval Latin theories of animal rationality.[4]

The setting of these theories was not restricted to commentaries on Aristotle. There are also many other sources in which one can find relevant contributions. These include, for instance, theological *Summae* and *Sentences* commentaries. They differ from commentaries on Aristotle with regard to content and function. Nonetheless, they must receive some attention in this context because later medieval theories of animal rationality are patchwork rather than all of a piece. This is illustrated by the fact that there is not a single medieval treatise on this subject. Rather, the question of animal cognition in general and of animal rationality in particular came up in very different kinds of texts, many of which do not, prima facie, seem to be concerned with these topics. But since nonhuman animals often functioned as a sort of heuristic tool in the Middle Ages (see Chapter 4), the discussion on their capacities was not limited to one particular genre but spread across various texts and disciplines which is why any reconstruction of late medieval theories of animal rationality must rely on various sources.

The sources of the present study are basically of two different kinds. The majority are texts which exist in critical Latin editions, such as many of the commentaries on Aristotle's *De anima* or Peter Lombard's *Sentences*. Some of these texts have already been translated into English or other European languages, either partly or entirely.[5] This also applies to some texts for which only uncritical Latin editions are available, such as Albertus Magnus' comprehensive *De animalibus*. The second type of source are those texts which have not yet been edited and are only available in early prints or manuscripts. A good example is Pseudo-Peter of Spain's commentary on *De animalibus* which, so

---

3 See Hasse (2000).

4 For a general overview of animals in Islamic philosophy and theology see Lauzi (2012), Wannenmacher (2017), and Adamson (forthcoming). On particular issues and authors see, for instance, Kruk (1995), (1997), (2001), and (2002); López-Farjeat (2010), (2012), and (2016); Druart (2016).

5 Unless otherwise indicated, all translations in this study are mine.

far, has received only very little attention.[6] These kinds of sources will not be used as extensively as sources of the first type but they will be used in order to establish a relatively broad basis which will help to show that there was not just one, but a whole spectrum of arguments and explanations in the later medieval period.

Despite the variety of sources, the present book does not aim to prepare a complete map of the theoretical landscape in the later Middle Ages. Instead, it aims to philosophically reconstruct, analyse, evaluate, and compare a wide range of views on four aspects of this problem: (1) universal cognition and concept formation, (2) judging, (3) reasoning, and (4) prudence (see Chapter 3 more below). Such a systematic approach does, of course, bear the risk of ignoring many aspects. For much could be said about the institutional context or the theological background of medieval theories. Yet, the advantage of this kind of approach is that it can deliver interesting philosophical insights into a historical discussion. This does not mean that we are to ignore its historical peculiarities. There are, of course, many significant differences between past and present theories of animal rationality, but there are also many astonishing parallels.

These continuities and discontinuities will be fleshed out by discussing the views of more than ten late medieval thinkers, namely, Albertus Magnus, Roger Bacon, Pseudo-Peter of Spain, Bonaventure, Thomas Aquinas, Giles of Rome, Peter of John Olivi, John Duns Scotus, William of Ockham, Gregory of Rimini, Adam Wodeham, John Buridan, and Nicole Oresme.[7] Some of these figures have already been widely read and studied, while relatively little is known about others. Moreover, some of them were Dominicans, others were Franciscans, and some were not affiliated with a particular order at all. Still, all of them did, to differing degrees, participate in the academic life of their time, mainly at Oxford and Paris. They were trained in philosophy and theology, commented Aristotle's writings as well as those of other authoritative figures, and all of them developed interesting – and different – views on various questions concerning animal rationality. This is an important finding, insofar as all of them subscribed to the metaphysical premise according to which only human animals are endowed with intellect and reason. That is, from a common starting-point they went in divergent directions, and the aim of this book is to track their paths.

6 I am grateful to Theodor W. Köhler for providing me with his transcription of this work.

7 This study's conception thus follows, for instance, Perler (1992), (2004) and (2011), or Pasnau (1997).

Tracking their paths does not, however, require one to write a history of the development of their theories. Whether the views of the early Ockham or the early Aquinas accord with those of the late Ockham or the late Aquinas is not the primary concern here. Just as little is the aim to examine whether there is a notable difference between the theories from the beginning of the second half of the thirteenth and those from the end of the first half of the fourteenth century. Instead, the aim is to depict the variety of theories that existed between 1250 and 1350. So far, only very little attention has been paid to this variety because the denial of reason to animals did not seem to leave many theoretical options open. That this assumption needs to be revised is the main thesis of this book.

CHAPTER 3

Structure and Key Questions

Later medieval theories of animal rationality were part of a larger discourse on (nonhuman) animals. It is important to understand this background before discussing these theories in more detail. Therefore, Part 1 provides some information on the role animals played in the Middle Ages, especially in medieval philosophy (Chapter 4). Briefly speaking, their philosophical role was to function as a sort of litmus test. This function derived from the common view of the metaphysical status of animals and, more precisely, from the usual and widely accepted distinction between the sensory souls they have and the intellectual souls of human beings. On the one hand, this psychological distinction drew a clear line between humans and other animals. Any kind of capacity found both in human and in nonhuman animals was easily identified as a capacity of the sensory (part of the) soul. To this belonged, for instance, simple perceptions but also memory (Chapter 5). By contrast, the cognition of universals or the formation of concepts, judging, and reasoning were usually taken to form the *triad of intellectual operations* (Chapter 6). On the other hand, this distinction was put to the test by the observation of comparatively complex, human-like behaviour in nonhuman animals. The question was: Could such behaviour be brought about by the powers of the sensory soul or must one ascribe intellectual powers to nonhuman animals? Accordingly, despite the seemingly clear boundary, there were obvious *grey areas* (Chapter 7). But how did authors of the thirteenth and fourteenth centuries deal with these grey areas?

The main parts of this study will address this question by looking at the three main intellectual operations, namely, *universal cognition and concept formation* (Part 2), *judging* (Part 3), and *reasoning* (Part 4). Although it was commonly held that nonhuman animals cannot engage in any of these operations, several questions arose with respect to each of them. For instance, do sheep see wolves or grey, furry patches? Put differently, do they see something *as* something – for instance, as something dangerous – like we do by means of concepts or is their perception of the world entirely different from our way of seeing things? How do nonhuman animals (mentally) represent things? Do they only represent this or that particular thing or can they also form some general representations, such as of water, for example, when looking for a drink? Furthermore, are their reactions to the stimuli they perceive based on acts of judgment? Do sheep flee wolves because they judge that wolves are dangerous? And if they do, how is this compatible with the definition of judgments

 | DOI 10.1163/9789004363779_005

as propositions? In modern terms, shall we ascribe propositional attitudes to other animals? And shall we go even further and grant them the capacity to reason, too? For how else could we explain their capacity for problem solving? When we deal with novel situations we think or reason about what would be the best thing to do, so why should not other animals handle such situations in the same way? It is these and other questions that will be discussed in Parts 2 to 4, which look at different late medieval answers given to them.

Part 5 then turns to what medieval thinkers called 'prudence' (*prudentia*). Prudence is slightly different from capacities like judging or reasoning, at least insofar as it is not part of the triad of intellectual operations. However, it is often mentioned in connection with nonhuman animals' cognitive capacities. In a way, one could say that prudence is (or was) the medieval analogue of our modern notion of animal intelligence because even those authors who clearly denied any kind of rational operations to other animals did not refrain from calling them prudent. But what does it mean to say that an animal is prudent or behaves prudently? According to the common definition which medieval authors adopted from ancient thinkers, acting prudently means to do something in the present by memory of past events and with foresight of future events. Many animals, such as ants, seem to do this when storing food for the winter months. But do they really have something like memory and foresight? And how do these compare to the human capacity to remember and plan? Are other animals as prudent as we are or shall we say that calling them 'prudent' is just metaphorical speech? Like in Parts 2 to 4, it will become clear that different answers were given to these questions.

The variety of answers is what then leads to Part 6. This final part will offer a closer comparison of later medieval and contemporary theories (Chapters 30 and 31) as well as an evaluation of the various explanatory strategies seen in Parts 2 to 5 (Chapters 32 and 33). The evaluation draws on a distinction that is often made in modern (animal) philosophy, namely, the distinction between *differentialism* and *assimilationism*. Usually, differentialists are taken to argue that there is a *fundamental* difference between us and other animals. Assimilationists, in contrast, are said to hold the view that this difference is only a matter of *degree*. In their opinion, we are much more similar to other animals than the differentialist perspectives leads us to believe. Now, prima facie, all medieval authors must be classified as differentialists because by denying the faculties of intellect and reason to nonhuman animals they obviously establish a fundamental or *anthropological difference*. Yet, if one classifies all medieval theories in this way, one risks losing sight of their variety. Hence, it is worth looking at what unites and what distinguishes the different theories of animal rationality. As it turns out, most, if not all late medieval authors agreed on a *metaphysical* differentialism. Nonetheless, they chose different strategies for dealing with this metaphysical premise and so they differed from a *methodological* point of view.

This is an important point also with regard to the comparison between medieval and modern theories of animal rationality because most modern scholars are inclined to see a sharp contrast between modern and medieval theories. Quite often, the latter are reduced to the denial of reason without asking what actually follows from this denial. Does this denial of rational *powers* or *faculties* automatically lead to the denial of rationality, that is, to the denial of rational *processes* or *operations*? Or did at least some medieval authors allow for what might be called 'rationality without reason'? In other words, did they deny *faculty rationality* but ascribe *process rationality* to nonhuman animals? As will be shown, there are at least some authors who can be read in this way. But even for those who did not go into this direction, there is an important parallel in comparison to contemporary theories of animal rationality: they all struggled with the question of how to define the *cognitive processes* that underlie a certain *behaviour*. This question remains an issue in modern research. And so even if the present book is unlikely to advance contemporary theories of animal rationality, it will hopefully elucidate some of the fundamental philosophical problems that unite past and present theories.

PART 1

Animals and Rationality in the Middle Ages

Introduction to Part 1

Later medieval theories of animal rationality arose in a particular historical context. It is important to understand some of the particularities of this context because, despite many similarities, animals and rationality did not play exactly the same role they do today. The first chapter of this part (Chapter 4) therefore sets out to explore the role of animals in the Middle Ages both in general and in philosophy particularly. Their philosophical role depended especially on the metaphysical status of nonhuman animals which was different from the status granted to humans: while both were regarded as having souls with vegetative functions and sensory powers, humans were seen to surpass other beings through their possession of intellect and reason. The nature and operations of these different types of soul will be summarised in Chapters 5 and 6. Particular emphasis will be placed upon the triad of intellectual operations. Although these operations seem to establish a clear dividing line between the souls of human and nonhuman animals, there are certain grey areas, as Chapter 7 shall show. These grey areas arise from the fact that there are many cases in which nonhuman animals behave in a very human-like manner and thus seem to have human-like cognition. This does not immediately compromise the animal/human boundary. However, it calls for an explanation, and so the grey areas are the source from which originate theories of animal rationality.

 | DOI 10.1163/9789004363779_006

CHAPTER 4

The Role of Animals in the Middle Ages

In the medieval world, nonhuman animals were almost omnipresent. They were essential for the functioning of daily life as they supplied transportation, food, and raw-materials, among other things. They lived at the courts of wealthy kings as well as in the stables of poor farmers. Thus, medieval people had regular contact with all kinds of animals, and their life would have been much more difficult, if not impossible, without the existence and support of various domestic and wild species.[1] Animals also played a prominent role in medieval art and architecture. They populated (and still populate) many manuscripts, paintings, and buildings, especially the portals of Gothic cathedrals, which overflow not only with biblical figures, saints, and demons but also with numerous animals, both realistic and fantastic.[2]

Yet, the religious role they had amounted to more than just decoration. In countless texts, they figure prominently as symbols for the wisdom and power of the divine creator whose 'marks' (*vestigia*) they bear, as Basil of Caesarea puts it in his *Homiliae in Hexaemeron*, which were well-known also to late medieval writers.[3] In a way, animals were thus considered evidence for what is nowadays called 'intelligent design'. Their physiological and psychological makeup seemed to be created relative to their needs, a consonance which, like everything else in the world, was taken to be a sign of God's intelligence. They were also signs in the sense that their physique and behaviour were interpreted as having allegorical meaning.[4] In this connection, they were often used as moral role models. Many of their behaviours were taken to be instructive and

1 On animals in daily life see, for instance, Dinzelbacher (2000), 181–211; Kiser (2007); Pascua (2007); Smets & Van den Abeele (2007); Obermaier (2009), 3–8; Salisbury (2011), 10–60. For a concise introduction to animals in medieval culture see Resl (2007a).

2 See Klingender (1971), Hicks (1993), and Resl (2007b) as well as the contributions in Flores (1996).

3 Basil of Caesarea, *Homiliae IX in Hexaemeron*, lib. IX, c. 4, ed. Migne (1886), 198B: "Ex his liquet, quo in omnibus inest natura, id doctrina non comparari, nihilque in rebus esse aut inordinatum, aut improvidum: sed res omnes sapientiae conditoris sui vestigia prae se ferre […]."

4 On animals as signs and symbols or '*vestigia Dei*' see Resnick & Kitchell (1999), 23–26; Beullens (2007); Page (2007).

 | DOI 10.1163/9789004363779_007

exemplifications of 'natural virtues' (*virtutes naturales*), such as the diligence of ants or the care of storks, were praised as worthy of imitation.[5]

Although one might think that this symbolic value only mattered in religious contexts, a look at so-called bestiaries and encyclopedias from the medieval period reveals that such symbolic views were woven into more zoological accounts of animals. Texts on animals in twelfth- and thirteenth-centuries encyclopedias and bestiaries usually combine zoological information with theological exegesis. This is hardly surprising because, first, they heavily relied on earlier writings, such as the *Physiologus* and Isidor of Seville's *Etymologiae* both of which also amalgamate zoology and theology.[6] Second, any zoological (or scientific) enquiry concerning earthly creatures was taken finally to lead to a better understanding of the divine creator. This idea is already present in Augustine, who presumably coined the metaphor of the 'book of nature'. According to this metaphor, God is not only the origin of Scripture but has also written the 'book of creation' (*liber creaturae*) which, of course, includes animals.[7] In order to acquire knowledge about God, one needs to carefully study both books he has 'written'. From a modern perspective, this might seem like a baneful submission of 'natural science' (*scientia naturalis*) and its subdisciplines, including zoology, to the rule of theology. However, one should not conflate the theological framing of this science with its contents. Of course, any scientific enquiry into nature was considered to bring about some theological surplus. But this does not mean that the *scientia naturalis* was entirely theologised.

In fact, in the twelfth century in particular, the discovery of the theological value of natural science led to, or at least facilitated, what Marie-Dominique Chenu called "the discovery of nature."[8] This discovery was part of a more general "Renaissance" which, besides an increased interest in ancient science and sources, also produced an "increasing intellectual and practical interest among educated people, especially in the universities, in studying the things of nature through physical causes understood rationally, apart from their spiritual potential," as Willene B. Clark put it.[9] This study of nature did not deny

5 See Salisbury (2011), 81–107; Toivanen (forthcoming); Oelze (forthcoming a).

6 On animals in medieval encyclopedias see Schmidtke (1968) and (1993); Hünemörder (2003); Obermaier 2009, 13–16; Friedrich (2015). On bestiaries see Orlandi (1985); Clark & McCunn (1989); George & Yapp (1991); Clark (1995) and (2006); Hassig (1995); Voisenet (2000); White (2002).

7 On this metaphor see Speer (1995), 30f., and Kann (2003), 35f.

8 Chenu (1968), 4–24.

9 Clark (2006), 2. On the "Renaissance" of the twelfth century see Chenu (1968) and Speer (1995).

its 'spiritual potential' by any means. But nature (as the object of natural science) was 'desacralised' to some extent.[10] For the purposes of the present study we need not go into the details of this twelfth-century "Renaissance." What matters, however, is that it set the stage for many of the discussions on the animal/human boundary which shall be examined in the subsequent parts. Since nature not only includes things like minerals and plants but also human and nonhuman animals, the increased interest in nature also led to a more intense study of humans and other animals.

In the thirteenth century, the comparison between human and nonhuman species even became a "philosophical and anthropological key paradigm," in the opinion of some scholars.[11] There were various reasons for this paradigm shift. Among the most important factors is the transmission, translation, and reception of the *Corpus Aristotelicum*. The translation of Aristotle's writings from Greek and Arabic into Latin had already begun during the first half of the twelfth century but it was not finished until the beginning of the second half of the thirteenth century.[12] It had a huge impact on the academic discourse in the Latin West. Not only did it enlarge the textual basis of medieval learning and teaching, but it also set a certain agenda. The main subjects of Aristotle's texts quickly became the dominant subjects in many fields, from logic and philosophy of science to physics, metaphysics, and ethics.

As regards animals, the so-called 'Books on Animals' (*De animalibus libri*) became particularly relevant. These books are in fact a collection of three separate writings on the history, the parts, and the generation of human and nonhuman animals (*Historia animalium, De partibus animalium, De generatione animalium*). They were, however, treated as a unified work by the Arabic translator(s).[13] Michael Scot who, presumably by 1220, had produced the first Latin translation, followed their lead on this point and so did William of Moerbeke when also translating these books into Latin around fifty years later.[14] Even though the *De animalibus* did not receive as much attention from Latin commentators as other texts of Aristotle, it formed an integral part of the (later) medieval discourse on animals. With his extensive commentary of

10 See Chenu (1968), 14.

11 Köhler (2001). For a thorough analysis of the comparison of human and nonhuman animals in the thirteenth century see Köhler (2008), 233–340. See also Köhler (2000), 52–73.

12 On the medieval translations and reception of Aristotle's writings see Dod (1982) and Lohr (1982).

13 On the Arabic translations see briefly Kruk (2003).

14 See Dod (1982), 48f., and, in more detail, Drossaart-Lulofs (1992).

the same title and the additional *Quaestiones super De animalibus*, Albertus Magnus in particular made a major contribution to this discussion.[15]

Although Aristotle's original treatise and Albert's and other medieval scholars' commentaries cover virtually every aspect of zoology, the different books of *De animalibus* were not the only Aristotelian treatises dealing with animals. There also were two other works, namely, *De motu animalium* and *De incessu animalium* (known as *De progressu animalium* in the medieval period) which concern the physiology and, to some extent, also the psychology of animals.[16] It was these writings which, together with the *De animalibus*, formed the medieval 'science of animals', the *scientia de animalibus*, in the broadest sense. Like today's zoology, which is a sub-discipline of biology, this science was usually considered to be a sub-discipline of the much broader 'natural science', the *scientia naturalis*, because animals are a type of natural body (or thing), as Thomas Aquinas famously observed.[17]

Yet, the science of animals did not particularly focus on the aspect of cognition. There certainly are various passages, both in Aristotle's books on animals and in the medieval commentaries and supplements, which contain statements on animals' cognitive capacities. But the more important and influential contributions to this field were Aristotle's writings on psychology, most importantly, his *De anima*.[18] In this work, he examines the souls of plants as well as nonhuman and human animals, and his theory gave rise to many debates. Since the *De anima* concerns not only the nature but also the operations of the soul (or souls), it is one of the prime works on psychology in general and on animal cognition in particular. Still, it is not the only place to look.

---

15 Besides Albert, there was, for instance, Peter of Spain who wrote a commentary on *De animalibus*, a critical edition of which has been published by Francisca Navarro Sánchez. On these commentaries in particular and the discourse on animals in general see Asúa (1991), (1994), (1997), and (1999) as well as Resnick & Kitchell (1999), 34–42. See also the contributions to Steel, Guldentops & Beullens (1999), especially, Köhler (1999). In addition, see Perfetti (2004). On some fourteenth-century debates in particular see Cova (1997).

16 On the Latin reception of these works see De Leemans (1999) and (2000). On the (late) medieval reception of Aristotle's theory of animal motion see also Frampton (2008).

17 Thomas Aquinas, *In primum librum Posteriorum analyticorum Aristoteles expositio*, c. 13, lec. 25, ed. Leonina I (1882), 237: "Sed intelligendum est unam scientiam esse sub altera dupliciter. Uno modo, quando subiectum unius scientiae est species subiecti superioris scientiae; sicut animal est species corporis naturalis, et ideo scientia de animalibus est sub scientia naturali."

18 On the extensive commentary tradition on the *De anima* see Bakker & Thijssen (2007); De Boer (2013).

Some minor treatises that were part of the so-called *Parva naturalia* also examine particular aspects of psychology, such as sensation (like the *De sensu et sensibilibus*) or memory (like *De memoria et reminiscentia*).[19] Moreover, many other writings which are not (primarily) concerned with psychology, such as the *Metaphysics* and the *Politics*, contain crucial statements on the cognitive capacities of animals. Hence, (later) medieval theories of animal cognition in general and of animal rationality in particular not only drew on those of Aristotle's writings that focused exclusively on psychological themes, but also upon other parts of the *Corpus Aristotelicum*, from treatises on zoology and psychology to works on entirely different topics.

Aristotle's works were, of course, not the only source for the formation of theories in the Latin West, especially in the field of psychology. As is well known, the influence of Arabic psychology should not be underestimated in this respect. Besides Averroes' commentaries on Aristotle's *De anima*, there is Avicenna's original *Liber de anima* in particular which found its way into Latin theories.[20] Arabic authors were well acquainted with Aristotle's writings, including his treatises on zoology and psychology, but many of them did not content themselves with critically discussing Aristotle's ideas. Rather, they also developed their own original theories, among which is Avicenna's highly influential theory of inner senses. This theory was of utmost importance for the discussion on animal rationality, as we shall see in the following chapter, because it clearly defined a number of corporeal faculties apart from the five external senses that humans share with other animals. This theory of inner senses complemented Aristotle's theory of the soul in a crucial regard.

However, neither in Aristotle's nor in Avicenna's psychology were nonhuman animals a subject in their own right. They were the subject of psychology, first and foremost, because like plants and humans they were ensouled beings and so the science of the soul would have to study all different kinds of souls rather than only one particular type.[21] Yet their souls were included in psychology for more than just the sake of completeness. Rather, their cognitive capacities played a pivotal heuristic role with regard to the study of human souls since any kind of enquiry into the nature of human souls needs to answer the question of whether human souls are unique and, if so, what it is that makes them

19 On the Latin reception of the *Parva naturalia* see the contributions to Crisciani, Lambertini & Martorelli (2004), especially, Federici Vescovini (2004). On the reception of *De memoria et reminiscentia* in particular see Bloch (2007), 137–228.

20 On the reception of Avicenna's *Liber de anima* see Hasse (2000).

21 Note that there was some debate over the subject matter of the '*scientia de anima*', see Köhler (2000), 335–368; De Boer (2013), 71–77.

unique.[22] This is the reason why there was an active interest in comparing the souls of humans and of nonhuman animals with each other. This comparison was not restricted to psychology. In almost all fields of medieval philosophy and science, nonhuman animals were used as 'a tool of knowledge' ("un outil de connaissance"), as Jacques Voisenet put it.[23] This means that, quite often, they were not the *explanandum*, that is, the thing which is to be explained, but the *explanans*, that is, the thing by which something else is explained.

A passage that nicely illustrates this explanatory role of animals is found in John Duns Scotus' *Quaestiones super secundum et tertium De anima*.[24] There Scotus discusses whether one should posit a so-called common sense, that is, a faculty which is superior to the particular senses, such as the sense of sight or taste. Scotus answers this question in the affirmative and his answer relies heavily on the psychological makeup of nonhuman animals. Like human beings, they do not only perceive things such as a piece of food but also experience pleasure (*delectatio*) in what they perceive. Feeling pleasure, however, is a cognitive act in which a power reflects on a particular object. When one enjoys the sweet flavour of a piece of chocolate cake it is not the sense of taste itself that creates the feeling of pleasure because what the sense of taste does is simply to receive the information that what it tastes is sweet. It is not up to the sense of taste to decide whether sweet flavour is pleasurable. If this is true, then the question is whose task it is to do this. One possible answer would be that it is the intellect's task to decide whether a certain perception is pleasing. Yet this answer is proven wrong when we look at nonhuman animals, Scotus argues. Like us they also enjoy certain things while they dislike others. According to the Aristotelian theory of the soul, however, they lack intellect and reason and so their pleasure must derive from a sensory power.

Regardless of whether Scotus' argument is convincing, it shows how animals were used in a particular philosophical context. They served as a kind of litmus test: in Scotus' argument they help to show that a certain power is sensory

22 See Toivanen (2013b), 357.

23 Voisenet (2000), esp. 329–352. Similarly, Köhler (2008), 234 and 237, states that nonhuman animals had the function of mutually elucidating various aspects ("wechselseitige Erhellungsfunktion"). Obermaier (2009), 2, calls them 'a medium of knowledge and visualisation' ("Medium der Erkenntnis und Vergegenwärtigung"). Davids (2017), 6, adopts Courtenay's idea of nonhuman animals as an 'analytical tool' ("analytisches Hilfsmittel").

24 See John Duns Scotus, *Quaestiones super secundum et tertium De anima*, q. 9, §10, eds. Bazán et al. (2006), 72f. According to the editors, this text is "a series of independently disputed questions concerning the soul, written down in connection with some teaching activity in the liberal arts" rather than a commentary on Aristotle's *De anima*; see ibid., 139*.

rather than intellectual and they do this in many other arguments, several of which we shall encounter in what follows. What is common to all of these arguments is that they rest on a very basic but crucial assumption: nonhuman animals have sensory souls by virtue of which they can engage in a variety of cognitive operations (see Chapter 5 below). There are, however, certain kinds of cognition in which they cannot engage because they require intellectual or rational faculties (see Chapter 6). Consequently, the kind of argument Scotus and many others present works only as long as this basic assumption about the animal/human boundary holds true. If there is a case in which this assumption is put to the test, namely, the case of animal rationality (see Chapter 7), it will have serious consequences for the arguments that rest on it.

CHAPTER 5

Animal Souls and Sensory Cognition

Among medieval philosophers, there was not much of a debate on whether dogs, cats, apes, or horses, and even more minute nonhuman animals, such as flies and bees, have souls. By and large, they all adopted the view of Aristotle according to which the soul is the 'principle of life'. Hence, all living beings, from plants to humans, possess souls; otherwise they would not be alive. For ordinary people, the idea that nonhuman animals have souls seemed to be an odd idea. As Roger Bacon states in the *Communia naturalium*, the 'crowd of laymen believes that only humans have souls which is why they laugh about the clergymen who say that dogs and other nonhuman animals have souls'.[1] Similarly, Adelard of Bath, in his twelfth-century *Quaestiones naturales*, includes his nephew's statement that the question of whether nonhuman animals have souls is 'ambiguous to men of our time'. Adelard, however, who is his nephew's interlocutor in this dialogue, stresses that only ordinary people would reply by clearly denying that animals have souls. Those, in contrast, who are familiar with natural philosophy do not at all hesitate to ascribe souls to other animals because they acknowledge that animals have 'sensation' (*sensus*), and sensation is a clear sign for the presence of a soul.[2]

Sensation or sensory cognition is indeed what characterises animal souls, and we shall see shortly what it means to have sensation. But before that it is important to note that the presence of sensation is only a sufficient, not a necessary condition for the ascription of a soul. The reason for this is that, despite the lack of any kind of cognition, plants have souls too, according to Aristotle's widely-accepted theory: trees and flowers nourish themselves, they grow, and propagate, and so they have what was usually called a *vegetative* soul. Admittedly, this type of soul does not enable them to see their food or to feel their partners, for example. But it endows them with the basic functions

1 Roger Bacon, *Communia naturalium*, lib. I, pars 4, c. 1, ed. Steele (1911), 283: "Immo vulgus laicorum in multis regnis adhuc credit quod soli homines animas habent, unde derident clericos qui dicunt canes et cetera bruta habere animas."

2 Adelard of Bath, *Quaestiones naturales*, q. 13, ed. and tr. Burnett (1998), 110–112: "NEPOS: His dissertis, postremo urtrum bruta animalia animas habeant necne, solvendum suscipe. Id enim quidem etatis hominibus ambiguum est. ADELARDUS: Ut vulgus de negatione non dubitat, ita philosophus affirmatio certa est. Habent enim, et eas habere sic assero. Bruta sensus habent."

 | DOI 10.1163/9789004363779_008

of life: nutrition, growth, and reproduction.[3] Since plants share these basic functions with nonhuman and human animals there is a certain continuity between them. This "principle of psychological continuity," as it is sometimes called, is best expressed by the idea of the 'great chain of being' or the 'ladder of nature' (*scala naturae*).[4]

According to this picture, all (living) beings occupy the different steps of a ladder. And since the single steps build upon each other – it is much closer to a stairway than a ladder – there is continuity insofar as what is found at one step will also be found at the next step. For instance, nutrition, growth, and reproduction are found in plants, nonhuman, and human animals alike. The latter two, in turn, share not only the functions of the vegetative soul but they also have in common the functions of the so-called *sensory* soul. This is actually what sets them apart from plants and thus establishes a certain *discontinuity*, for plants count as living or ensouled beings, but unlike dogs or humans they are not animals.

In order to qualify as an animal, a living being needs to have sensation in addition to vegetative functions. In a nutshell, 'animals' (*animalia*) are living beings that have a sensory soul. In this sense, the concept 'animal' is a generic term which applies to various species of animal, both human and nonhuman.[5] Thus, it establishes some sort of 'generic community' between various kinds of animals.[6] It is certainly true that several authors tried to make a terminological distinction between human and nonhuman animals, for instance, by referring to the latter as 'irrational animals' (*animalia irrationalia*), 'brutes' (*bruta*), or 'beasts' (*bestiae*).[7] Many of these expressions are pejorative rather than neutral and intend to emphasise the inferiority of nonhuman animals. Nonetheless, they do not exclude dogs, horses, or ants from the realm of animals. They only show that this realm can be further divided in one way or another. Since this was commonplace among medieval philosophers, to speak of 'humans and *other* animals' or 'human and *non*human animals' is not to take an ideological stance towards the status of nonhuman animals. For medieval thinkers humans

---

3 On the history of plant souls see Ingensiep (2001).

4 On the principle of psychological continuity see Matthews (1978). On the idea of a great chain of beings see Lovejoy (1936), esp. 24–98. On the Aristotelian *scala naturae* see Granger (1985) and Coles (1997).

5 There are even some exceptional cases in which celestial bodies and demons were taken to be animals by ancient and medieval thinkers, see Köhler (2014), 105f.; Sander (2016), esp. 262.

6 Köhler (2008), 184. Similarly, Davids (2017), 31, claims that it establishes a certain 'animal conformity' ("animalische Konvenienz").

7 See Resl (1997), 3–10; Preece (2005), 1; Köhler (2008), 226–232; Salisbury (2011), 10–12. In earlier medieval texts one can also find terms like '*belua*' (for wild animals) or '*pecora*' (for cattle).

clearly are living beings that have a sensory soul, and this is what unites them with all other kinds of animals.[8] But what does it actually mean to have a sensory soul? What is it that makes humans and other animals stand above plants on the *scala naturae*?

Above all, it is the possession of a number of organs by virtue of which they can engage in certain activities or operations or by virtue of which they have certain capacities. Of course, the possession of organs which have certain functions is nothing specific to sensory souls. The functions of the vegetative souls also depend on various organs: there are organs for ingestion, digestion, and for reproduction. What matters is rather that the organs of the sensory soul are *other* organs than those of the vegetative soul and so they can bring about *other* operations. As mentioned before, one can summarise the operations of sensory souls as *cognitive* operations. This means that they are "mechanisms by which animals acquire, process, store, and act on information from the environment," to quote a definition of cognition that has become widely accepted in contemporary philosophy and psychology.[9] By and large, this definition also catches the medieval notion of cognition.[10] Still, it is important to note that medieval thinkers used to distinguish between sensory and intellectual cognition. One might therefore prefer to speak of 'perception' when talking about sensory operations. However, most of the authors covered by this study did not refrain from calling perception a kind of cognition (*cognitio*).[11] Hence, it is neither anachronistic to call perception a kind of cognition nor is it wrong to claim that human *and* nonhuman animals have cognition because both of them clearly have various organs for sensory cognition.

Regarding the organs of sensory cognition a distinction was usually made between two different kinds of senses, namely, *external* senses and *internal* senses. The external senses are five: sight, hearing, smell, taste, and touch. All of them have corresponding organs: eyes, ears, nose, tongue, and skin or flesh.[12]

8 This point is also emphasised by Resl (2007a), 3, who says that "[...] *animal* was used in its strictest Latin sense to refer to all breathing, moving, living beings, that is, to humans and nonhuman animals alike. In this language system no single word was available that corresponded to our modern animal in referring to all nonhuman animals. [...] modern scholars who neglect this difference can all too easily jump to anachronistic conclusions." On the medieval concept, see also König-Pralong (2011).

9 Shettleworth (2010), 4.

10 See Pasnau (1997), 4.

11 On scholastic theories of perception see, for instance, Knuutila (2008); Hasse (2010); Smith (2010); Perler (2015c).

12 In *De anima* II.7-11, Aristotle discusses at length the senses and their corresponding organs.

Yet, if one compares the external senses of human and nonhuman animals, one will easily see that there are at least three important differences between the senses of different species. The first difference is that not all animals have the same *number* of external senses. While humans usually have all of these senses, there are other animals that lack certain of them. Bees, for instance, do not possess the sense of hearing, in the opinion of many late medieval thinkers, such as Thomas Aquinas. Even though bees are obviously frightened by loud noises and so seem to hear, this is only because they feel 'the motion of the air' (*commotio aeris*) that is produced by such noises, Aquinas explains.[13] Lack of hearing is the reason why bees cannot be taught because 'hearing is the sense of instruction' (*auditus est sensus disciplinae*), as Aquinas remarks in the same place with reference to Aristotle's *De sensu et sensato*. However, they are not seriously impaired by this lack because they still manage to find nectar and pollen and the combs they built are of an admirable geometrical regularity.[14] Thus, one can be an animal without having all of the five senses. This is even more obvious with regard to those species that do have only one external sense. A prominent example is oysters, whose only external sense is touch. Therefore, they are only slightly superior to plants.[15] Nonetheless, they qualify as *animals* because, in contrast to plants, they have at least one sense and hence can engage in sensory cognition.

Besides differences in number between the external senses of human and nonhuman animals there are also differences in *physiology*, that is, in the corporeal quality of the sense organs. The eyes of bees have a different structure than human eyes, for example, and similar differences occur in the physiology of any other of the four senses: the ears of elephants are much larger than human ears and so are the noses of horses. Although these differences might seem to be mere differences of shape or structure, they actually point to a third and more fundamental difference, namely, a difference in the way in which the various external senses function. In other words, there is a difference in the senses' *degree of perfection*. Certain species, such as hawks, for instance, surpass humans with respect to sight as they are capable of spotting

13 Thomas Aquinas, *In duodecim libros Metaphysicorum Aristotelis expositio*, lib. I, lec. 1, n. 12, eds. Cathala & Spiazzi (1964), 8. On this see passage also Davids (2017), 173f.

14 On the medieval *topos* of the sagacity of bees see Guldentops (1999) and Roling (2013b), esp. 412–418.

15 See, for instance, Thomas Aquinas, *Summa contra gentiles*, lib. II, c. 68, ed. Leonina XIII (1918), 440: "[…] sicut quaedam infima in genere animalium parum excedunt vitam plantarum, sicut ostrea, quae sunt immobilia, et solum tactum habent, et terrae in modum plantarum adstringuntur."

prey over large distances. The reason for this is that their eyes have a different 'complexion' (*complexio*), as Albertus Magnus explains in his *De animalibus*.[16] And so from physiological differences result various functional differences. In many cases, nonhuman animals benefit from these differences in comparison to humans, insofar as they have more powerful external senses than we do. Recognition of this sensory superiority has been a commonplace among philosophers since Antiquity. In his *De anima*, Aristotle already mentions that our sense of smell is much worse than the sense of smell of many other animals, and in medieval encyclopedias dogs are often mentioned as an example of this superiority.[17] Other examples are the sight of lynxes, the taste of monkeys, or the touch of spiders.[18] In sum, there is a large variety among animal species with regard to their external senses.

However, perception via the external senses is not the only kind of sensory cognition. Beyond that, many animals can, for instance, also imagine or remember things they have sensed. But imagination and memory do not take place at the level of the external senses: if one imagines or remembers a rose, one does not presently see, smell, or touch a rose. Instead, one can, for instance, close one's eyes and still have a mental image of a rose. Therefore, imagination and memory were considered to be so-called 'inner senses' (*sensus interiores*). The theory of internal senses was extremely influential in the later Middle Ages but its history is highly complicated as there was much disagreement about various aspects.[19] There was, first, disagreement over how many inner senses there are at all and, in particular, over how many of these are shared by human and nonhuman animals. In most cases, four inner senses are mentioned: (i) common sense, (ii) imagination or phantasy, (iii) estimation, and (iv) memory. In addition, many authors also mention a

---

16 Albertus Magnus, *De animalibus*, lib. XXI, tr. 1, c. 1, ed. Stadler (1920), 1323.

17 See Aristotle, *De anima* II.9, 421a7-13. On the dog's sense of smell see, for instance, Isidor of Seville, *Etymologiae*, lib. XII, c. 2, §25, ed. André (1986), 111.

18 These examples were often summarised in a proverb which can be found, for instance, in Thomas of Cantimpré, *Liber de natura rerum* IV.1, ed. Boese (1973), 106: "Homo in quinque sensibus superatur a multis: aquile et linces clarius cernunt, vultures sagacius odorantur, simia subtilius gustat, aranea citius tangit; liquidius audiunt talpe vel aper silvaticus: Nos aper auditu, linx visu, simia gustu, Vultur odoratu precedit, aranea tactu." On ancient versions of this saying see Sorabji (1993b), 15f. On thirteenth-century views of the differences between human and nonhuman animals' external senses see also Köhler (2014), 248–266.

19 On the medieval theory of inner senses and its history see Wolfson (1935); Klubertanz (1952), esp. 37–151; Harvey (1975); Kemp & Fletcher (1993); Federici Vescovino, Sorge & Vinti (2005); Hasse (2010); Kärkkäinen (2011); Knuutila & Kärkkäinen (2014).

cogitative power, though for some this was an additional power in humans, whereas others took this to be the human analogue of nonhuman animals' estimative power. Similarly, some held that imagination and phantasy are one faculty, while others considered them to be different powers.[20] Some authors also argued that there is actually only one inner sense which has different functions.[21] Besides disagreement on the number of internal senses there was also much debate over where the single senses are located in the brain.[22] And third, there was a wide range of views on how these senses function, especially in different species of animals.[23]

All of these differences are more than mere technicalities because the way in which one describes the inner senses determines one's view of the cognitive capacities of animals. For the moment, it suffices to highlight three points that are particularly noteworthy in this regard. First, just as some animals lack certain external senses, so do some lack certain internal senses. Flies, shellfish, and worms, for instance, do not have memory according to Albertus Magnus.[24] Flies regularly return after we have chased them off because they are incapable of remembering our hurtful blows. Dogs, in contrast, have an outstanding memory. A popular example among ancient and medieval authors is Argos, the dog of Ulysses, who after many years of separation was the only one to recognise his master after his return.[25] Thus, there are not only differences in number but also differences in the degree of perfection of the inner senses and the memory of dogs is just one example. As regards imagination, this difference was an important criterion for differentiating between 'perfect' and 'imperfect animals' (*animalia perfecta* and *imperfecta*).[26] While the former are capable of imagining a rose without presently perceiving a rose, the latter can only have a bundle of sensory impressions of a rose in their imagination when

---

20 Lisska (2016), 219, nicely calls this discussion "a historical muddle." There was also much lamentation among medieval thinkers about this, as Köhler (2014), 266f., shows. See also Hamesse (1988).

21 This position was, for instance, held by Peter of John Olivi; see Toivanen (2013a). A similar view is found in Buridan; see De Boer (2014) and Lagerlund (2017).

22 On theories of brain function in particular see Sudhoff (1913) and Clark & Dewhurst (1996), 8–53.

23 A comprehensive overview of thirteenth-century positions is provided by Köhler (2014), 266–374.

24 Albertus Magnus, *De animalibus*, lib. XXI, tr. 1, c. 2, ed. Stadler (1920), 1326; *De memoria et reminiscentia*, tr. 1, c. 1, ed. Donati (2017), 115; *Metaphysica*, lib. I, tr. 1, c. 6, ed. Geyer (1960), 8.

25 On this example see Roling (2011), 223.

26 See Köhler (2006) and (2014), esp. 329–340.

presently perceiving a rose. As in the case of the external senses, these differences were mostly explained by differences in the quality of the bodily organs, that is, the specific parts of the brain in which the inner senses reside. Consequently, whether one has a good memory or a bad imagination depends on the bodily mixture of certain parts of the brain, and much as there are more or less slight differences between the individuals of one species, there are more or less profound differences between the inner senses of one species and another.[27]

A third point that is particularly relevant for theories of animal rationality is the role of one specific inner sense, namely, the so-called 'estimative power' (*vis/virtus aestimativa*) or 'estimation' (*aestimatio*). This power is definitely one of the, if not *the*, most crucial inner sense of nonhuman animals because its function is to judge what is perceived. For instance, it is the power that determines whether the thing perceived is harmful or harmless. The estimative faculty somehow tells the organism what it should do in response to a certain perception. The classical example of how this power works is found in Avicenna, whom most scholars credit with inventing the theory of estimation.[28] In his *Liber de anima*, Avicenna introduces the estimative power by referring to the example of a sheep seeing a wolf. The sheep, Avicenna says, not only perceives the sensory 'forms' (*formae*) of the wolf, such as the wolf's colour and shape, its smell, or sound. In addition, it also perceives what, in the Latin, was rendered as 'intention' (*intentio*), namely, the wolf's harmfulness.[29] It is this intention of harmfulness which then triggers the sheep's reaction of flight. And so the estimative power is essential to the sheep's survival because without this faculty no reaction would be triggered.[30]

One might, of course, immediately ask whether the estimative power triggers nothing but *instinctual* reactions because the sheep's reaction of flight is clearly an innate response that can be found in all sheep. But this question is a pivotal part of many of the discussions to be considered more closely in

---

27 An exemplary and detailed discussion of these differences is provided by Albertus Magnus, *De animalibus*, lib. VIII, tr. 6, c. 1, ed. Stadler (1916), 668–670.

28 For Hasse (2010), 314, this theory marks "the most successful addition to Aristotle's faculty theory." On Avicenna's theory of estimation see further Black (1993) and (2000), 59–62; Hasse (2000), 127–153; Hall (2006); Pormann (2013), 102–107; López-Farjeat (2016), 63–66.

29 Note that '*intentio*' is a technical term which is usually rendered as 'intention' although is has little to do with intentions in the modern sense of the term because the sheep does not literally grasp that the wolf *intends* to be harmful; see Crane (1998), esp. 816, and Amerini (2011), esp. 558f.

30 See Avicenna Latinus, *Liber de anima seu Sextus de naturalibus*, lib. I, c. 5, ed. van Riet (1972), 86–89, and lib. IV, c. 3, ed. van Riet (1968), 34–40. For a discussion of this example and its reception in the Latin West see Piro (2005); Perler (2006) and (2012b).

Parts 2 to 5. For now, it is enough to say that *all* animals, no matter whether they are human or nonhuman and no matter whether they are big or small, have sensory souls. This means that they possess a number of external and internal senses, even though these senses differ in number, physiology, functionality, and so forth. In every case, they provide *cognitive* access to the world and thus go beyond the basic functions of life provided by the vegetative soul. Sensory souls surely have much in common with vegetative souls. Most prominently, their operations are bound to bodily organs. For instance, the sense of sight does not work without eyes, and the faculty of memory is situated in a certain part of the brain where various sensory impressions can be stored, just as digestion requires an intestinal tract. This corporeal foundation of sensory cognition is the reason why sensory souls are often characterised as *material* souls: they are material in the sense that having a sensory soul means to have certain *bodily* organs. This corporeal nature of the senses is common to both human and nonhuman animals and so we share many things with dogs, horses, bees, and other animals.[31]

31 As the chart in De Leemans & Klemm (2007), 158, shows, we actually share the *majority* of powers with other animals according to the Aristotelian theory of the soul.

CHAPTER 6

Human Souls and the Triad of Intellectual Operations

As we have seen, all animals are superior to plants in that they have not only vegetative powers but also sensory, that is, *cognitive* powers. The possession of cognition is what establishes a psychological continuity between nonhuman and human animals. However, there is also a crucial discontinuity and it is this discontinuity that brings us closer to theories of animal rationality. In short, the discontinuity between humans and other animals consists in the fact that only the former have *rational* or *intellectual* souls. Consequently, humans are not simply animals but a very peculiar kind of animals, namely, 'rational animals' (*animalia rationalia*).[1] They are, so to speak, 'animals *plus x*' with 'x' being the faculties of intellect and reason. Nonhuman animals are, in turn, 'animals *without x*'. This lack of intellect and reason is why they were often called 'irrational animals' (*animalia irrationalia*) or 'brute animals' (*animalia bruta*), as mentioned before. The possession of a rational soul thus establishes what modern scholars usually call an *anthropological difference* because it sets humans apart from all other animals.

Medieval philosophers stressed this point. In his commentary on *De animalibus*, Peter of Spain, for instance, states that 'humans excel any other animal by the nobility of intellect and reason' (*homo enim excellit quodlibet animal nobilitate intellectus et rationis*).[2] In his *Sentences* commentary, Peter of John Olivi defines human beings as 'intellectual animals or animals having intellect' (*bestiae intellectuales seu intellectum habentes*).[3] On the one hand, this shows that the animal nature of human beings was not denied. On the other hand, it emphasises that something crucial is added to this animal nature, namely, intellectual or rational faculties. Consequently, if these faculties are subtracted from a human animal, one no longer has a *human* animal. All that remains is an

1 '*Animal rationale*' is simply the Latin translation of Aristotle's Greek '*zoon logon echon*' from *Politica* I.2, 1253a7-10.

2 Peter of Spain, *Questiones super libro 'De animalibus' Aristotelis*, lib. 1, q. 4, ed. Navarro Sánchez (2015), 117. See also Köhler (2008), 191.

3 Peter of John Olivi, *Quaestiones in secundum librum Sententiarum*, q. 57, ed. Jansen (1924), 338.

 | DOI 10.1163/9789004363779_009

animal, as Henry of Ghent points out in his *Quaestiones quodlibetales*.[4] Therefore, it is obviously the rational soul that defines humans qua humans and so 'in reality, mankind is nothing else but the rational soul', as Thomas Aquinas puts it in the *Summa contra gentiles*.[5] Since this definition of humans was usually taken for granted by medieval philosophers, one can rightly say that it was a fundamental ontological presupposition or a fundamental metaphysical and anthropological presupposition to which they subscribed.[6] Yet, the most interesting question is, how do the rational souls of humans compare to the sensory souls of nonhuman animals? If having a soul means having certain capacities or engaging in certain operations that are characteristic of this soul, what are the specific operations of rational souls?

In order to answer these questions one might first need to ask whether there is actually a difference between 'intellect' (*intellectus*) and 'reason' (*ratio*) because, on the one hand, they seem to be used interchangeably, and yet, on the other hand, it would be superfluous to speak of intellect *and* reason if they were not distinct powers. This question seemed to puzzle medieval authors, too. In the *Summa theologiae*, Thomas Aquinas, for example, explicitly addresses the question of whether intellect and reason are 'separate powers' (*diversae potentiae*).[7] He first presents a number of opinions from other authors who seem to suggest that intellect and reason are indeed distinct powers. The author of the influential *De spiritu et anima*, for instance, says that there is a clear hierarchy: first come the external senses, then the internal senses, then comes reason and then comes the intellect. Hence, the intellect is superior to reason. Augustine, in contrast, employs them synonymously and so do others.

Aquinas' tries to solve this contradiction by arguing that intellect and reason are not separate powers. Nonetheless, he concedes that there is a certain difference between rational and intellectual cognition, between '*ratiocinari*' and '*intelligere*'. While the latter means 'to apprehend an intelligible truth per se' (*simpliciter veritatem intelligibilem apprehendere*), the former means 'to proceed from one intelligible object to another in order to cognise an intelligible truth' (*procedere de uno intellecto ad aliud, ad veritatem intelligibilem cognoscendam*). One could illustrate this difference by comparing it to the difference between solving a mathematical problem by intensive thinking or by

4 Henry of Ghent, *Quaestiones quodlibetales* I, q. 15, ed. Macken (1979), 93: "Excepto enim eo quod intellectus est, non manet homo nisi bestia [...]."

5 Thomas Aquinas, *Summa contra gentiles*, lib. IV, c. 81, ed. Leonina XV (1930), 253: "Et secundum hoc, humanitas non est aliud realiter quam anima rationalis."

6 This is how Köhler (1992), 718, and (2008), 295, puts it.

7 See Thomas Aquinas, *Summa theologiae* I, q. 79, a. 8, ed. Leonina V (1889), 274f.

sudden inspiration. If one solves the problem by intensive thinking it takes some time until one arrives at the solution. One might have to search for the right approach, take notes, and so forth. That is, one makes various steps and so jumps from one partial solution to another. In the case of a sudden inspiration, however, the solution immediately appears. One sees the problem and – in a flash of insight – knows the solution. The latter is certainly a much more impressive way of solving mathematical problems. In this sense, *intelligere* is somehow superior to *ratiocinari*. Still, both ways lead to the same result and both cognitive acts are brought about by one and the same cognitive power, according to Aquinas.

Admittedly, Aquinas' view is only one among many. Other medieval authors had (slightly) different thoughts about the distinction between reason and intellect.[8] For the present study, however, the more decisive question is whether this distinction matters for the aspect of animal rationality. On the one hand, it seems to be negligible because neither of these faculties of the soul was ascribed to nonhuman animals by any of the authors covered here. So even if some of them held the view that there is a difference between the two, it remains relatively insignificant if one is looking at animals which, by definition, lack both of these powers. On the other hand, the distinction might not be entirely irrelevant insofar as there were at least some authors who ascribed the capacity of 'reasoning' (*ratiocinari*) to other animals, as we will see in Part 4. Thus, the slight inferiority of rational cognition in comparison to intellectual cognition might actually imply a greater proximity to the powers and operations of the sensory soul (see Chapter 33). But before dealing with these details we first need to see what kind of operations and capacities were usually taken to come along with the possession of a 'rational' or 'intellective soul' (*anima rationalis*; *anima intellectiva*). What are the typical tasks or operations of the higher faculties?

In most accounts there are three main intellectual operations, namely, (i) universal cognition and concept formation, (ii) judging, and (iii) reasoning. These operations built upon each other, as, once again, Thomas Aquinas points out.[9] First of all, the intellect 'apprehends the quiddity of a thing' (*apprehendit quidditatem rei*), say, the quiddity of a rose. Apprehending the quiddity of a rose significantly differs from apprehending a rose via the senses.

8 For the full picture from Augustine to Buridan see Enders (2001); Speer (2001); Hoenen (2001). On the Aristotelian and Neoplatonic origins of the distinction see also briefly Sorabji (1993b), 74f.

9 See Thomas Aquinas, *Summa theologiae* I, q. 85, a. 5 co., ed. Leonina V (1889), 341. On this article see also the analysis by Pasnau (2002), 273.

If one sees, smells, or touches a rose, for example, one perceives certain features of a *particular* thing. It is *this* or *that* rose one sees or smells or touches and one would not have any of these perceptions if no rose were *presently* there. Hence, sensory cognition is of present and particular things. But what if one remembers or imagines a rose? In this case, one does not presently perceive a rose but still has a sensory impression. In fact, one can have sensory impressions without *presently* perceiving things. But these sensory impressions are nonetheless the impressions of a *particular* thing such as of the particular rose one saw yesterday in the garden.

To apprehend the quiddity of a rose, however, means to *abstract* from any material and temporal conditions because the quiddity of a rose is what is common to *all* roses, not only to this or that particular rose one has perceived at this or that particular time and place. The colour red, for instance, does not belong to the quiddity of roses because there are also white or yellow roses. Having petals, on the other hand, seems to be part of the quiddity of roses because all roses have petals even if they are coloured differently. Once one has grasped the defining universal features of roses, one knows the universal 'rose'. One could also say that one has formed the concept 'rose'. This concept is universal in that one can apply it to all roses no matter where they stand, what colour or size they have, when one has seen them, and so forth. Since this kind of cognition goes beyond sensation, it was usually attributed to the faculties of intellect and reason.

There are, admittedly, many medieval theories of how concepts are formed, and, likewise, there are numerous theories of what universals are.[10] However, all of these theories suggest that the cognition of universals and the formation of concepts rely largely, if not entirely, on the powers of the rational soul. Many authors stressed that the sensory faculties play a role in this process insofar as they deliver the necessary material out of which concepts are formed. That is to say, we would be incapable of forming the general term 'rose' without having previously sensed particular roses. Nonetheless, it is the intellect that cognises the universal and forms the concept 'rose'. Consequently, the intellect is also responsible for the second operation, the task of 'composing or dividing' (*componere vel dividere*) these concepts, that is, the formation of judgments.[11] Once we have acquired concepts, we can combine them with others concepts *ad libitum*. We can, for instance, form propositions such as 'Roses are plants' or 'Roses are a genus in the family of *Rosaceae*'. Thus, general terms or concepts are the building-blocks of judgments or propositions, at least according to the

10 For an overview see, for instance, Adams (1982) and Holopainen (2014).

11 See Thomas Aquinas, *Summa theologiae* I, q. 85, a. 5 co., ed. Leonina V (1889), 341.

so-called "'building-block model' of propositions and concepts."[12] This model is still influential today and it was, in one way or another, held by the majority of late medieval authors, too.[13]

According to the building-block model, it does not actually matter what kind of theory one holds concerning the formation of concepts because concept formation is a different cognitive act than the formation of propositions. Therefore, someone like William of Ockham who held a different theory of concept formation than Aquinas nevertheless agreed that 'to form propositions and to syllogise is the proper task of the intellect' (*formare autem propositiones et syllogizare proprium est intellectui*).[14] As Ockham's phrasing shows, this not only applies to the *formation* of propositions, that is, the act of judging, but also to the act of 'syllogising' (*syllogizare*). Aquinas adopts a slightly different terminology by saying that the third intellectual operation consists in 'going from one composition or division to another, which means to reason' (*ex una compositione vel divisione ad aliam procedere, quod est ratiocinari*).[15] Still, syllogising and reasoning are synonymous insofar as syllogising is a prominent form of reasoning. If, for instance, we move from the premises 'All roses are plants' and 'All plants have a vegetative soul' to the conclusion that 'All roses have vegetative souls', we have syllogised as well as reasoned because our reasoning took the form of a syllogism, more precisely, a syllogism of the so-called *modus Barbara*. We have thus combined a certain number of propositions or judgments – the premises – in order to arrive (by way of dividing and combining their concepts) at another judgment – the conclusion. Such reasoning, then, is the third element of the triad of intellectual operations.

These intellectual operations amount to more than just a mere play with words. They are, as Albertus Magnus points out, the foundation of human knowledge because they lead us to 'the principles of arts and sciences' (*principia artium et scientiarum*).[16] Thus, one could say that humans are rational animals not only because they form and combine concepts and propositions but also because they form a whole body of knowledge which they divide into different fields or disciplines. This is also one of the reasons why Aristotle's

---

12 On this model see, for instance, Glock (2010), 15–19.

13 For a brief overview of medieval theories of judging see Tachau (1993) and Perälä (2014). The most comprehensive and insightful studies are Nuchelmans (1973) and (1980); Perler (1990) and (1992).

14 William of Ockham, *Quaestiones in IV Sententiarum* (*Report.*), q. 14, eds. Gál & Green (1984), 288.

15 See p. 39 n11.

16 Albertus Magnus, *De animalibus*, lib. XXI, tr. 1, c. 2, ed. Stadler (1920), 1328.

logical writings played such an important role in medieval philosophy and other disciplines. His logic – or the medieval teaching of his logic, more precisely – begins with the study of *concepts* (in *De interpretatione*), continues with the study of *judgments* (in the *Categoriae*), and ends with the study of *syllogisms* (in the *Analyticae* as well as in the *Topica* and the *De sophisticis elenchiis*). It thus provides the foundation for and guarantees the validity of any other discipline because any mistake in the division or combination of concepts or propositions will produce erroneous conclusions.

One could, of course, argue that sensory cognition also produces some kind of knowledge. If I see that this petal is red, I have also acquired some knowledge about it, at least in a broader sense of the term. Still, even if one accepts such a broader notion of knowledge, there remains a difference between the knowledge deriving from the senses and the knowledge created by reason and intellect: while the former remains at the level of particulars, the latter is universal. This also applies to statements about particulars. If one forms the judgment 'This rose is a flower', one does, on the one hand, make a statement about a particular rose. On the other hand, however, one employs general terms, such as 'rose' or 'flower'. Similarly, when making statements about individuals, such as Socrates, one employs general terms because one says, for instance, that 'Socrates is a human being'.

This universality is characteristic of intellectual cognition, and it was usually explained by the peculiar nature of the intellect. Unlike the powers of the vegetative and the sensory soul, the intellect has no organ, as Aristotle famously says in *De anima* III.4.[17] Unlike the eyes or the intestinal tract, intellect and reason are not corporeal or material powers. Rather, they are *immaterial*. For many medieval interpreters of Aristotle, this "immateriality condition" is a necessary condition with a view to the three operations of concept formation, judging, and reasoning because they all *abstract* from material conditions.[18] Whereas the sensory impression of a rose is material, at least to some extent, the concept 'rose' is not.

Admittedly, this is an overly simplified account of the intellect's immateriality. Many of the medieval theories were much more sophisticated than this account suggests and there was much debate over how this immateriality is to be understood.[19] Furthermore, there were also some authors, such as

17 Aristotle, *De anima* III.4, 429a19-28.

18 I borrow the expression "immateriality condition" from Black (2009), 331.

19 On some of the Islamic, Jewish, and Christian theories of the intellect see Black (2009).

John Buridan, who doubted that there is a (necessary) connection between the intellect's operations and its nature.[20] However, this was the view of a minority. The majority of authors agreed that intellect and reason are immaterial. For Christian authors, this view was also attractive because it was highly compatible with the doctrine of the *immortality* of the human soul. For if one supposes that the human soul is immaterial, it is much easier to explain how it can persist after the death of the body. Again, the details of this doctrine gave rise to many discussions. And, especially before the thirteenth century, there were also some authors, such as John Scotus Eriugena and Adelard of Bath, who held not only that human and angelic souls are immortal but the souls of nonhuman animals, as well.[21] Yet, as in the case of the intellect's immateriality, this was not the view of a majority.

The connection between the intellect's *operations* and its *nature* shows that it is not the case that the animal/human boundary can only be described in terms of *capacities*. It can also be described in more general terms as being largely identical with other dividing lines. First, it is identical with the *sensory/intellectual* divide because nonhuman animals have sensory powers while humans have intellectual faculties in addition. Since the latter were usually taken to be immaterial, the animal/human boundary is, second, co-extensive with the divide between *material* and *immaterial* powers. Of course, even sensory souls are immaterial insofar as they are forms, not matter. And yet, sensory faculties are bound to corporeal organs and it is in this sense that they are material. If one takes for granted that only immaterial powers can provide cognitive access to universals, the animal/human boundary runs, third, along the lines of *particularity* and *universality*. Furthermore, immaterial powers survive the death of the body and so there is a fourth dividing line between humans and other animals, namely, the line that divides *mortal* from *immortal* souls. Humans, of course, stand on both sides of these divides. They have both sensory *and* intellectual, corporeal *and* incorporeal powers, hence access to both particulars *and* universals. Moreover, some of their powers pass

---

20 On Buridan's position see Klein (2016).

21 See John Scotus Eriugena, *Periphyseon*, lib. III, c. 39, ed. Jeauneau (1999), 169–172; Adelard of Bath, *Quaestiones naturales* q. 14, ed. Burnett (1998), 118. On Eriugena's view see Nitschke (1967), 240–242; Dronke (1985), 817–822; Lauzi (2012), 208–217. The idea that nonhuman animals' souls are immortal can also be found in later authors such as Nicholas of Cusa; see Dohm (2013). Furthermore, it is present in medieval Islamic thought; see Kruk (1995), 31, and Druart (2016), 73.

away at the moment of death, while some persist.[22] This standing on both sides is why humans are a species of animals, not angels.[23] Nonetheless, they are a very peculiar species of animals because they significantly differ from other animals in various regards.

---

22 This idea gave rise to numerous discussions on the cognitive capacities of so-called 'separate souls' (*animae separatae*); see, for instance, Roling (2015).

23 On human cognition in comparison to angelic cognition see Roling (2008) and Iribarren & Lenz (2016), esp. Part III. On animals, humans, and angels, see Daston (2005).

CHAPTER 7

Grey Areas

There are obvious differences between the souls of humans and the souls of nonhuman animals, both with regard to their nature and with regard to their capacities. The cognition brought about by material sensory powers is different from the cognition brought about by immaterial intellectual powers. The latter can engage in concept formation, judging, and reasoning, while the former cannot. An animal that sees a red rose could thus be either human or nonhuman. A honey bee flying through the garden and looking for nectar has certain sensory impressions of roses just as we have certain sensory impressions of roses when passing by a rose. The bee, however, does not possess the concept 'rose'. It is capable of distinguishing roses from stones and other things but, according to this theory, it does not conceptualise its perceptions. Moreover, it forms neither propositions nor syllogisms because syllogisms consist of propositions and propositions are made of concepts. Therefore, a bee is not a rational animal because it does not engage in *rational cognitive processes* due to the lack of *rational faculties*.

Like every theory, this theory does its job when providing satisfactory explanations for certain phenomena. For instance, if one asks why an animal X exhibits a certain behaviour B, the obvious answer would be that X has certain (cognitive) faculties F which enable her to engage in certain (cognitive) processes P that bring about B. If B were a behaviour that cannot be satisfactorily explained without reference to inferential processes, one would ascribe the faculties of intellect and reason to X because it is these faculties in virtue of which living beings can engage in reasoning. Within the Aristotelian theory of the soul, this ascription is entirely sound as long as X is a *human* animal. But what if X is a *nonhuman* animal? If a nonhuman animal shows a kind of behaviour that seems to be based on intellectual processes – and we shall come across many examples of such behaviours in the following parts – it puts the above-mentioned definition of the animal/human boundary to the test. In other words, the animal's behaviour points to certain *grey areas*, that is, areas in which the boundary between the cognitive capacities of humans and other animals seems to be blurred.

Late medieval authors were well aware of the existence of such grey areas. For instance, in *De animalibus*, Albertus Magnus states that there are good reasons to cast doubt on nonhuman animals' lack of reason. He mentions the

 | DOI 10.1163/9789004363779_010

example of bees that get together in order to leave the hive with their king.[1] This they do, he says, 'by foresight for the community' (*provisione communitatis*) and 'not because of any individual advantage' (*non de utilitate propria*). This kind of care for the community is hardly possible 'without some spark of reason' (*sine aliqua scintilla rationis*), in Albert's view.[2] Similarly, there are many cases in which animals seem to have knowledge of how they can cure a certain disease from which they suffer, he says. They know, for instance, that a certain kind of plant is an appropriate remedy for snakebites. In other words, they have 'experiential knowledge' (*experimentum*), like a physician who prescribes aspirin against headaches. And although he cures individuals, his knowledge is universal because he knows that aspirin is likely to cure the headache not only of this particular patient but of other patients, too. Therefore, 'it is not likely that they [i.e., those nonhuman animals that know how to cure diseases] do not also somehow participate in reason' (*non videtur probabile, si non etiam utcumque ratione participant*).[3] Therefore, one can rightly have 'some doubt' (*aliqua dubitatio*) that all nonhuman animals lack reason.

A similar worry is expressed by Thomas Aquinas and Giles of Rome. In his commentary on *De sensu et sensato*, Aquinas points out that there are some people who have doubts that humans are the only animals with intellectual faculties. The reason is that higher-developed nonhuman animals, such as monkeys or apes (*symia*) especially, engage in certain actions or do certain works (*opera*) that are 'somehow similar to the works of reason' (*quedam similia operibus rationis*).[4] These behaviours make people question the common definition of the animal/human boundary. Giles of Rome also seems to know people who harbour similar qualms because in his commentary on *De anima* he states 'that there is doubt among some people on whether other animals, apart from humans, possess an intellect' (*quod dubium est apud quosdam, utrum alia animalia ab homine habeant intellectum*).[5] One of the sources of this doubt is the example of pygmies. Pygmies were usually classified as apes, that is, as 'non-humans' (*non-homines*). Nonetheless, they were sometimes taken to

1 Like the majority of ancient and medieval authors, Albert believed that bees have a king, not a queen.

2 Albertus Magnus, *De animalibus*, lib. VIII, tr. 6, c. 1, ed. Stadler (1916), 667.

3 Ibid.

4 Thomas Aquinas, *Sentencia libri De sensu et sensato*, tr. 2, c. 2, ed. Leonina XLV.2 (1985), 110: "Dicit autem [Aristoteles]: '*forte*', propter quosdam qui dubitaverunt de quibusdam aliis animalibus ab homine utrum habeant intellectum, propter opera quedam similia operibus rationis, sicut sunt opera symiarum et quorundam huiusmodi animalium."

5 See Köhler (2014), 106.

possess intellect and reason because of the human-like behaviours they show, as Giles explains.[6] Although Giles himself does not agree with this view, like many other authors of his time, he acknowledges that there are various behaviours seen in nonhuman animals which create difficulties for the Aristotelian theory of the soul.

Now there are several ways to deal with such difficulties. In order to see more clearly what the options are, it might help to schematise the problem as follows. The initial point is the presupposition or fundamental anthropological premise that nonhuman animals A_{nh} lack intellect and reason, that is, rational faculties F_r:

(P_0) All A_{nh} lack F_r.

Now we have an animal, X, and X is classified as nonhuman:

(P_1) X is A_{nh}.

Next comes the observation that X shows a certain behaviour B:

(P_2) X exhibits B.

B is a behaviour which is taken to be based on a rational process P_r, hence:

(P_3) B is based on P_r.

According to Aristotelian faculty psychology, P_r requires rational faculties F_r:

(P_4) Engaging in P_r requires F_r.

The premises P_2, P_3, and P_4 now lead to the following conclusion:

(C) X possesses F_r.

Given P_1, this conclusion obviously clashes with the fundamental anthropological premise P_0 because if *all* nonhuman animals lack rational faculties and if X is a nonhuman animal, it must lack intellect and reason like any other

6 Ibid. On the medieval disagreement over the status of pygmies see Koch (1931); Köhler (1992); Thijssen (1995); Roling (2010), esp. 491–498.

nonhuman animal. As said, there are, however, various ways to deal with this problem.

The first (and maybe most radical) solution to this contradiction would be the *revision of* P_0. Instead of saying that '*all* nonhuman animals lack rational faculties' one could simply say that '*some* or *most* nonhuman animals lack rational faculties'. This would allow for the existence of nonhuman animals which have rational faculties. One would, of course, need to specify *which* nonhuman animal species are endowed with such faculties but an obvious criterion for this distinction would be whether or not an animal shows behaviour B. If X shows B, it is endowed with intellect and reason.

If one is, for whatever reason, unwilling to ascribe intellect and reason to X, one can also choose another option, namely, the *revision of* P_1. In this case, one would not touch the fundamental anthropological premise. Rather, one would change the way of classifying X. If X is a pygmy, for example, one could say that pygmies are human, not nonhuman animals. This is less radical a solution since P_0 remains intact. All that changes is the classification of X based upon X's showing B.

A third possibility is to doubt that X exhibits B. That is, one could argue for a *revision of* P_2, or even for a rejection of this premise. The advantage of this option is that neither the fundamental anthropological premise nor the classification of X would need to be revised. But this option might come at the price of rejecting certain facts, and so it might be untenable, especially if one observes X repeatedly exhibiting B. Therefore, it seems more attractive to choose between two other options, namely, the *revision of* P_3 or the *revision of* P_4.

Revising P_3 means to question the argument from analogy. As mentioned in Chapter 1, the argument from analogy claims that from similar *effects* we can infer to similar *causes*. In the context of psychology, effects are *behaviours* and causes are *processes* and *faculties*.[7] A revision of P_3 would cast doubt on this argument insofar as it would introduce an alternative explanation for the *causes* of some behaviour B. Take the following example from modern research in comparative psychology: A chimpanzee and a human child solve the so-called 'floating peanut task'. They thus manage to figure out that filling water into a narrow empty tube at the bottom of which lies a peanut will bring the peanut within reach as the water level rises.[8] Now one might be inclined to say that this kind of problem solving requires reasoning (and reasoning is a rational process, hence requires rational faculties, a medieval philosopher would add).

7 The causes are processes *and* faculties because, according to faculty psychology, processes are based on faculties. Therefore, one could say that faculties are *primary* causes of behaviour.

8 See Hanus et al. (2011).

But one could also argue that it does *not* require reasoning. It might simply be based on a very fundamental process of association or it might be instinctual. This would solve the above-mentioned problem because if X's behaviour B is not based on a rational process P_r then X's exhibiting B does not lead to the conclusion that X is endowed with rational faculties F_r. The advantage of this option is that P_0, P_1, P_2, and P_4 remain intact. Still, one now has to assume the burden of explaining how B is brought about, if not by a rational process.

If one is unwilling (or unable) to assume this burden, one can choose another option. Instead of revising P_3 one could also revise P_4. According to P_4, the rational process P_r upon which the behaviour B seems to be based requires rational faculties F_r. Yet one might say that P_r does *not* require F_r. However, by unfastening the connection between P_r and F_r one has to face the challenge of explaining how other faculties, such as the faculties of the sensory soul, can bring about a rational process. Even if one manages to do this, one still has to face either of the following consequences. The first would be a degradation of the rational process P_r. For instance, reasoning would lose its status as an exclusively rational process because it would no longer take place in a rational faculty. Another consequence would be the upgrading of those faculties where P_r takes place. If this place corresponds to any of the inner senses, for example, the sense in question would be upgraded insofar as it would be able to engage in a process that was usually taken to require higher, that is, rational faculties.

As will become clear in the following parts, late medieval authors considered all of these options, but the majority of them favoured the latter two. They were relatively reluctant to revise the fundamental anthropological premise P_0 but they were quite open to discussing the connection between *behaviour*, *processes*, and *faculties*. This means, first, that they discussed whether a rational process such as reasoning is necessary for bringing about a certain behaviour or whether an alternative, non-rational process, can bring about this behaviour, too. Second, if they took this process to be rational (or close to rational) they discussed whether it actually requires rational faculties or whether it could also be realised by sensory powers. And if they did *not* take this process to be rational, they tried to give an account of the analogous (cognitive) process that is responsible for this behaviour.[9] It is in this sense that they developed *theories of animal rationality* because they discussed what it means to be rational or, more precisely, what kind of behaviours require rational cognitive processes, such as concept formation or reasoning.

9 This strategy can be identified as 'differentialism', while the former can be called 'assimilationism', see Chapters 31 and 32.

Rationality, of course, amounts to more than being capable of conceptualising or reasoning. There are a plethora of capacities without which one seems to be incapable of engaging in any of these operations. One of the most obvious capacities is language, and language is actually one of the most frequently mentioned criteria in modern debates about the animal/human boundary. In contrast to other animals, humans seem to be the *animals that speak*.[10] However, what is distinctive about language other than the formation of propositions? Of course, one can *communicate* non-verbally, and this is what many nonhuman animals do by making certain sounds.[11] Still, one can hardly *speak* without forming and combining concepts. At least from the medieval point of view, the possession of intellect and reason is more fundamental than the possession of language. This also applies to other capacities because if they are based on more than sensory operations, intellect and reason come into play, as does the triad of intellectual operations. Thus, late medieval theories of animal rationality might not end with these operations, but they definitely begin with them.

Another point that should be noted is that all of these theories are, of course, embedded within a broader *theory of the soul*. Therefore, any answer given to questions concerning animal rationality somehow depends on the answer that is given to the following questions, some of which have already been touched on in previous chapters: Are vegetative, sensory, and rational souls different kinds of souls or are they different parts of one and the same soul? If the latter, how do they interact? Is there a difference between the sensory souls of humans and the sensory souls of nonhuman animals? Is there a difference between the sensory souls of different species of nonhuman animals? How does the lack or the possession of intellect and reason affect the operations of sensory powers? More generally, how does the nature or the metaphysical status of the soul affect its operations or capacities?[12]

Most, if not all, of these questions play a crucial role in theories of animal rationality because they establish the basic framework within which the latter are elaborated. Rather than discussing them separately we shall approach them only from the vantage point of animal rationality for two reasons. The first reason is that some of these questions might not actually have an immediate impact on judgments regarding the cognitive capacities of nonhuman animals.

10 For a modern defence of this position, called 'lingualism', see Barth (2010).

11 On medieval views on animal communication and language see Köhler (2014), 393–489. In addition see Eco et al. (1989); Resnick & Kitchell (1996); Rosier-Catach (2006); Roling (2013a).

12 On these questions see Köhler (2006), 1086, and (2008), 172f.; Zupko (2008), 172; Perler (2015b).

For instance, whether an author is a so-called 'unitarist' and thus claims that there is only one soul that has different parts or a so-called 'pluralist' and hence thinks that vegetative, sensory, and rational souls are distinct, makes a difference when it comes to questions such as 'Does the intellect grasp particulars, and if so how?' or 'How does the conceptualisation of perceptual content work if the rational soul is distinct from the sensory soul?' Still, these questions only occur with respect to human souls and human cognition, because in this case one needs to answer how the human sensory soul relates to the rational soul. In the case of nonhuman animals, however, this question does not occur because they do not have rational souls.

A question that arises is whether this is a plausible theory, that is, whether it can coherently account for the differences that exist between living beings. One might assume that this problem disappears once one discards the Aristotelian theory of the soul. And, indeed, one does not struggle with the question of what distinguishes sensory and rational souls if one holds that something like a soul does not exist at all. However, at least one fundamental problem always remains no matter what theory of the soul one adheres to, and this is a second reason for adopting such a perspective. This problem concerns the relation between cognition and behaviour. If one supposes that the behaviour of animals is (maybe not entirely, but at least partly) based on certain cognitive processes, then the question is what these cognitive processes are and how they differ between different species. This problem is what unites medieval and modern theories of animal rationality, although they start from comparatively different psychological theories and also rely on different methods. Medieval thinkers, for instance, put less weight on empirical observation of animal behaviour. Instead, they took many of their examples from the literature they used.[13] Nonetheless, they asked what brings about this behaviour and, like modern scholars, they gave different answers to this question, as we shall see in the following parts.

13 On the role of observation in comparison to literary knowledge see Köhler (2008), 267–272.

PART 2

Universal Cognition and Concept Formation

Introduction to Part 2

As the previous part has shown, medieval thinkers considered the boundary between nonhuman and human animals to be identical with the boundary between sensory and intellectual powers. As we have seen, this boundary is also identical, first, with the dividing lines between materiality and immateriality and, second, between particularity and universality. Accordingly, nonhuman animals stand on one side of the divide since they possess material senses by means of which they can perceive particulars. Humans, by contrast, stand on both sides of the divide because they are additionally endowed with immaterial intellectual powers which enable them to cognise universals. This, of course, has an impact on how different animals perceive or, as one could also say, on how they *mentally represent* the world. To give an example, a sheep only perceives this or that particular wolf. Humans, in contrast, also cognise the universal 'wolf'. In modern terms, we possess the concept 'wolf' by means of which we can refer to all particular wolves. The sheep, however, lacks such a concept. In medieval terms, one would say that it does not grasp the 'quiddity' (*quiditas*) of wolf or 'wolfness'.

There are countless medieval theories on what universals or concepts actually are, what kind of metaphysical status they have, and how they are cognised or acquired.[1] Still, in many, if not all, of these theories the intellect is assigned a crucial role when it comes to the question of how universals are cognised or of how concepts are formed. The intellect's prominent role usually derives from its immaterial nature because only an immaterial power was usually taken to be capable of abstracting universals from particulars. Therefore, those animals that lack an intellectual soul also lack the access to the realm of universals. By and large, one might find this argument rather unproblematic. For why should we ascribe universal cognition and concept formation to nonhuman animals, if we can convincingly account for their behaviour without such an ascription?

If one takes, for instance, Avicenna's famous and influential example of the sheep and the wolf (see Chapter 5), it seems that no universals or concepts are needed in order to explain the sheep's reaction of flight. Rather, according to Avicenna, two things are needed: first, a particular internal sense faculty, called 'estimation' or 'estimative power' (*aestimatio* or *vis aestimativa*), and second, a particular kind of object, called 'intention' (*intentio*). The perception of an intention by virtue of estimation causes a certain bodily reaction and this is exactly what happens in the case of the sheep seeing a wolf. By means of its

1 For an overview see, for instance, Adams (1982) and Holopainen (2014).

 | DOI 10.1163/9789004363779_011

external senses the sheep perceives the wolf's sensible 'forms' (*formae*), such as its colour, its shape, its smell, or its sound, but what makes it run away is the estimative perception of the wolf's intention of hostility.[2]

Admittedly, intentions are not clearly material objects, and their perception involves abstraction. The estimative power 'abstracts them from matter' (*abstrahit eas a materia*), as Avicenna says, and he also points out that, unlike sensible forms, intentions 'are in themselves non-material things' (*sunt in se res non materiales*).[3] Nevertheless, they are 'materialised' (*materiatae*), 'bound to the sensible form' (*ligata cum forma sensibili*) or 'mixed with sensibles' (*commixtae cum sensibilibus*).[4] They are 'insensible' (*insensibile*) only insofar as they cannot be perceived by the external senses. Still, estimation is an inner sense, that is, a material power with a specific location in the brain.[5] Therefore, the abstraction of intentions by virtue of estimation is not the same as the abstraction of universals by virtue of an immaterial intellect. The perception of the intention of hostility is different from acquiring the concept of hostility: while the former "is apprehended as tied to a particular perceptible form," as Richard Sorabji puts it, the latter is independent of the perception of a wolf, for example.[6] So, in sum, no universals or concepts seem to be involved in the cognitive process that triggers the sheep's reaction of flight.

Yet, one might object, this is only one possible way of telling the story of the sheep's cognition. And even if it is the correct way of telling it, there might still be other cases of animal behaviour in which universal cognition plays a role. Indeed, many modern interpreters of Avicenna, as well as medieval thinkers in the Latin West, wondered whether the cognition of sheep and other nonhuman animals really stays within the realm of particulars or whether one must ascribe the capacities of universal cognition and concept formation to them, too. The questions they raise and the arguments they make concern various aspects of the problem but can be reduced to at least four questions.

A first question is the *question of conceptualisation*. This question is based on the fact that – in modern discussions – conceptualisation is often defined as 'seeing some X *as* F'.[7] Now this can be said about the sheep and the wolf.

2 See Avicenna Latinus, *Liber de anima seu Sextus de naturalibus*, lib. I, c. 5, ed. van Riet (1972), 86–89, and lib. IV, c. 3, ed. van Riet (1968), 34–40.

3 Ibid., lib. II, c. 2, ed. van Riet (1972), 118f.

4 See ibid., 119, and lib. IV, c. 3, ed. van Riet (1968), 38. This aspect is also stressed by Perler (2012b), 35, and López-Farjeat (2016), 65.

5 See Avicenna Latinus, *Liber de anima*, lib. I, c. 5, ed. van Riet (1972), 89.

6 Sorabji (1993b), 64. On abstraction in Avicenna in general, see Hasse (2001) and D'Ancona (2008).

7 See, for instance, Allen & Hauser (1991), 227, and also Glock (2010), 27f.

Because instead of saying that the sheep perceives the wolf's intention of hostility we could also say that it perceives the wolf *as* dangerous or *as* an animal that is to be fled. Accordingly, it does not simply see a grey furry thing, but a wolf or a dangerous animal. Thus, the sheep conceptualises the content of its perception just like we do when we see wolves or sheep or trees or houses. Of course, it will be hard, if not impossible, to determine exactly *how* or as *what* the sheep conceptualises perceptual content. In other words, it is not clear how it mentally represents what it perceives. It might see a wolf or a dangerous animal or something else.[8] But what matters for the moment is mainly whether it conceptualises at all.

What is closely linked to the question of conceptualisation is, second, the *question of categorisation*. So if one wonders how a sheep sees the world, one might ask whether it is capable of categorising or classifying the particular things it perceives. There is no doubt that humans do this, because we constantly categorise and classify things: we classify ferries and cars as means of transportation or apples and bananas as fruits. But what about a sheep? Does it put wolves in the category of harmful things and fellow sheep in the class of harmless things? If it does, there is reason to believe that it employs universals or concepts to some extent because it perceives some individual X as falling under the *general category* F or under the *concept* F. As in the case of conceptualisation we do not, of course, know exactly *how* the sheep groups individual things together.[9] But, as in the previous case, what matters most is whether it categorises things at all.

A third question that arises is the *question of generalisation*. Evidently, the sheep not only flees one wolf in particular, but wolves in general. Therefore, one might argue that it generalises because from the hostile nature or appearance of *one* wolf it infers the hostility of *all* wolves. It thus engages in what Jonathan Bennett (following Mowrer) calls a "one-many generalization."[10] This question is even more evident with respect to another famous example, which is also mentioned by Avicenna, namely, the behaviour of a dog when it sees a stick or a stone.[11] This example differs from the example of the sheep insofar as, unlike wolves for sheep, sticks and stones are not natural enemies of dogs. Nevertheless, dogs begin to perceive them as such once they have been beaten

8 On this problem of indeterminacy of representations see Glock (1999), 179.

9 Again, see Glock (1999), 182.

10 See Bennett (1989), 104.

11 See Avicenna, *Liber de anima*, lib. IV, c. 3, ed. van Riet (1968), 39. This example is also used by later authors, such as Leibniz; see Perler (2009), 86.

with them. Since they flee not only the particular stick or stone they have been beaten with, but sticks and stones in general, they seem to generalise.[12]

With respect to all of the examples mentioned here, one can also raise, fourth, the *question of object identification*. How does the sheep, for instance, know that what it sees is a wolf or a harmful thing? Maybe, it possesses some sort of repository of common forms or images and, when it perceives a particular thing, identifies it as wolf, for example, by matching the form of this particular thing with the common form of wolf it has stored in its 'mind'. If this is how the process works, the sheep possesses not only mental representations of this or that particular wolf but also a more general mental representation of 'wolf'. The ascription of such general representations becomes even more plausible if one tries to explain how animals manage to find food or drink.[13] For instance, a horse which is hungry and looking for grass is going to eat *any* grass it finds. It does not only eat one particular portion of grass. Thus, it seems to look for grass in general and its search is likely to be based on a general, rather than on a singular, representation of grass.

As we have seen, there is a whole range of questions concerning universal cognition and concept formation in nonhuman animals. Before turning to the individual answers that were given to these questions by various thirteenth- and fourteenth-century authors, a couple of things should be noted. First, some of the questions that have been sketched above are tightly intertwined. For instance, the question of object identification is, in some sense, a reversal of the questions of conceptualisation and categorisation because in all cases there are two things involved in the cognitive process: one general, or universal, and one particular mental representation. The only thing that is different is the order of cognition, so to speak. Still, it would be wrong to reduce all of them to a single question because, as will become clear in what follows, they all cover slightly different aspects of the problem. However – and that is the second point worth noting – one needs to be careful with regard to how those questions are posed in the medieval texts. The way in which the questions have been phrased above is inspired by the way in which they occur in modern

12 In the modern discussion, generalisation seems to be taken to be less demanding than conceptualisation and categorisation, because even someone like Davidson, who denies concepts and propositional attitudes (e.g. beliefs and desires) to nonhuman animals, grants them the capacity of generalisation; see Davidson (1982), 326.

13 One can, of course, argue that one should avoid talking about (mental) representations at all, as does Glock (1999), 179. But even if one finds this argument generally convincing, it cannot be applied to medieval positions, because (most) medieval authors do take representations to exist, in one way or the other. On mental representations in medieval philosophy see, for instance, Panaccio (2010) and Klima (2015).

discussions of animal cognition as well as in the secondary literature on the medieval debate. Medieval authors, however, did not literally speak of conceptualisation, categorisation, generalisation, or object identification. Nevertheless, they did discuss the differences between the perception of the world of humans and nonhuman animals, and most of what they say can be seen as a reply to one or more of the above-mentioned questions.

In the following, we will look at the replies of Thomas Aquinas, Albertus Magnus, Pseudo-Peter of Spain, Roger Bacon, John Buridan, and Peter of John Olivi. The order is thus not chronological, but rather follows systematic criteria. While Aquinas and Albert are relatively hesitant to ascribe any kind of cognition to nonhuman animals that goes beyond the level of particulars, the other thinkers are more open in this respect. That is, they do at least consider some possibilities in which nonhuman animals might gain access to the level of universals. This is not to say, however, that Aquinas stands at one end of the spectrum, whereas Olivi stands at the other end. The differences are subtler. Furthermore, all of these thinkers have at least one thing in common: they all examine the way in which nonhuman animals perceive the world and they do this in order to find out whether the boundary between sensory and intellectual or material and immaterial powers is really the same as the boundary between particularity and universality.

CHAPTER 8

Estimation, Conceptualisation, and Categorisation (Thomas Aquinas)

One of the most prominent medieval texts concerning the question of how nonhuman animals perceive the world and whether they conceptualise and categorise what they perceive is undoubtedly a very short passage from Thomas Aquinas' *Sentencia libri De anima*. Although quite short, it has garnered an enormous amount of attention in the secondary literature. This makes it sometimes hard to disentangle Aquinas' own position from what is ascribed to him by modern interpreters. Of course, every claim about Aquinas' (or any thinker's) position is an interpretation in itself. So it might be impossible to determine what his 'own' view was. Nevertheless, it is worth trying to separate what he says from the various suggestions that have been made regarding how his words are to be understood. In order to better comprehend why Aquinas comes to talk about the cognitive capacities of nonhuman animals in his commentary on Aristotle's *De anima*, it will be helpful to briefly look at the context of his statements.

In the thirteenth lecture of the second book, Aquinas comments on the difference between so-called sensibles *per se* and sensibles *per accidens*. To illustrate the difference, he gives the following example. If one sees a white thing, the whiteness of this thing is a sensible *per se*. This is because white is a colour and colours are the proper sense objects of sight. Now it might accidentally be the case that the white thing one sees is also a sweet thing, say, a white chocolate bar. The sweetness is not a sensible *per se* for my sense of sight because sight only perceives colours. Nonetheless, one can perceive the sweetness by sight, namely, *per accidens*.[1] Thus, one can perceive things *per accidens* with one sense (e.g. sight) which have been perceived *per se* by another sense (e.g. taste). Obviously, it always takes at least two different cognitive powers to sense *per accidens*. Each of these powers perceives certain things *per se*, but each of them can also sense *per accidens* what the other has sensed *per se*. This kind of cognition is usually called 'accidental' or 'incidental cognition'.

Now the combination (or cooperation) of two external senses is only one possible way of bringing about accidental cognition. There are also other

1 See Thomas Aquinas, *In Aristotelis librum De anima commentarium*, lib. II, lec. 13, n. 395, ed. Pirotta (1959), 101.

 | DOI 10.1163/9789004363779_012

cognitive powers which allow us to perceive things *per accidens*. One of those powers is the intellect. If one sees something that is talking, Aquinas says, one also perceives – *per accidens* – that this thing is living because being alive is a necessary prerequisite for being able to talk and our intellect informs us about this connection.[2] Besides our intellect or another external sense, there is also a third cognitive power able to produce accidental cognition: the so-called 'cogitative power' (*vis cogitativa*). In human beings, this faculty is one of the four internal senses besides common sense, imagination, and memory. The cogitative power plays a crucial role in the conceptualisation of perceptual content as well as in categorisation because it enables us to perceive individuals as members of kinds or 'as existing under a common nature' (*ut existens sub natura communi*), as Aquinas puts it.[3] It makes us see *this man* or *this animal* and thus, in a way, brings together particulars and universals. To be clear, the cogitative power is a bodily power and thus still belongs to the realm of materiality and particularity. Nonetheless, it is 'united to the intellectual part of the soul' (*unitur intellectivae*) and thus somehow bridges the gap between the sensory and the intellectual by permitting us to observe that the particular things we see fall under certain universals.[4] So by virtue of the cogitative power we see the world the way we do, namely, as a collection of objects. Of course, one could describe what one sees as mere collection of coloured patches. But this is not the way in which we usually describe our perceptions. Rather, we say that we see objects, such as tables, trees, or books.[5] But what about the way in which nonhuman animals perceive the world? Do they also see tables, trees, or books?

According to Aquinas, nonhuman animals clearly lack the cogitative power. What they possess, however, is the estimative power (*vis aestimativa*). It is this power from which many accidental perceptions arise.[6] For instance, if a sheep

2 Ibid., n. 396.

3 Ibid., n. 398: "Nam cogitativa apprehendit individuum, ut existens sub natura communi; quod contingit ei, inquantum unitur intellectivae in eodem subiecto; unde cognoscit hunc hominem prout est hic homo, et hoc lignum prout est hoc lignum."

4 The most extensive study on the cogitative power in Aquinas is still Klubertanz (1952). More recent studies include Barker (2007); Di Martino (2008), esp. 85–101; Tellkamp (2012b); De Haan (2014). Concise introductions to the history of cogitation can be found in Summers (1987), 198–234, and Rubini (2015), 464–475.

5 For Lisska (2006), 12, this is what makes Aquinas' theory of perception stand out against the "bundle view paradigm" of the British empiricists because, in his view, Aquinas' theory better accounts for our experience of sensation which "is of things rather than of sense data."

6 This is why the theory of intentions can be interpreted as an Arabic 'response' to Aristotle's theory of sensibles *per accidens*, see Di Martino (2008), 14.

perceives the colours or sounds of a lamb, it might also perceive that what it hears or sees is its offspring. Thus, it sees a white thing *per se*, but *per accidens* it sees its offspring. Or it sees a grey, furry thing *per se* and a thing that is to be fled *per accidens*. In every case, this accidental perception is caused by the estimative power which apprehends individual intentions, according to Aquinas.[7] The question now is what this accidental perception by virtue of estimation amounts to. Does it make the sheep see lambs and wolves instead of white and grey patches? Moreover, does it make the sheep see individuals as members of kinds? In other words, is it capable of conceptualising and categorising perceptual contents?

In the secondary literature on Aquinas, there is a whole range of differing opinions regarding these questions. A first group of interpreters claims that Aquinas does indeed consider nonhuman animals to be capable of conceptualisation and categorisation. For instance, Judith Barad thinks that "in attributing apprehension to animals, Aquinas is, in effect, indicating that they can, to some extent, conceptualize."[8] Mark Barker claims that "[b]y the estimative, an animal categorizes the object as harmful."[9] And Paul Hoffman states that "while he [i.e., Aquinas] thinks animals do not conceive of universals as such, that is, apart from particulars, they do conceive of particulars as falling under universals."[10]

Other interpreters do not go that far but still hold that by way of grasping intentions "a sheep not only sees a grey patch of colour, but sees a threatening wolf," as A. Mark Smith puts it.[11] Cyrille Michon (adopting Michael Dummett's notion of 'proto-thoughts') suggests that nonhuman animals' estimative perceptions amount to "proto-concepts."[12] So, in Michon's view even if the sheep

7 See Thomas Aquinas, *In Aristotelis librum De anima commentarium*, lib. II, lec. 13, n. 397, ed. Pirotta (1959), 101: "In animali vero irrationali fit apprehensio intentionis individualis per aestimativam naturalem, secundum quod ovis per auditum vel visum cognoscit filium, vel aliquid huiusmodi."

8 Barad (1995), 117.

9 Barker (2012a), 204f. This paper is a revised part from Barker (2007) where he defends the general thesis that Aquinas' denial of recognition of kinds to nonhuman animals can be modified to allow for such a knowledge of kinds.

10 Hoffman (2012), 164f.

11 Smith (2010), 336f. Smith is actually speaking about Avicenna's theory of intentions but since Aquinas adopts Avicenna's theory, Smith's statement might also apply to Aquinas' account. A similar interpretation can be found in Tellkamp (1999), 172; (2006), 1355; and (2012b), 611.

12 Michon (2001), 333f. Kaukua (2014), 110, also argues that intentions are "quasi or proto-concepts."

does not possess concepts, such as 'wolf' or 'grass', it does at least see 'enemy' or 'food' when it sees a wolf or grass. For him, this is less then using full-fledged concepts, but still more than simply perceiving coloured patterns. Thus, nonhuman animals are endowed with "a complex kind of knowledge," and this knowledge somehow permeates their perception of the world, according to Jörg A. Tellkamp's reading of Aquinas.[13]

A third group of scholars is even more cautious. As Robert Pasnau remarks, "we cannot say – at least not strictly speaking – that the sheep senses that the object approaching is a wolf." The reason is that sensing wolves (instead of grey, furry somethings) requires the concept of 'wolf'. Hence, "to apprehend a wolf as a wolf requires the intellect" because only the intellect is capable of working with concepts.[14] Similarly, Eleonore Stump holds that "recognizing what one is perceiving depends on an act of the intellect," more precisely, on "the first operation of the intellect, namely, determining the quiddity or whatness of a thing."[15] Robert Pasnau hesitates to conclude that nonhuman animals do not see objects (which, in his view, is more demanding than seeing grey, furry things). But even if they do see objects, they "do not see them as members of kinds."[16] So, according to Pasnau's interpretation of Aquinas, "the sheep doesn't put the wolf into the category of dangerous things but is simply able to perceive danger at the same time as perceiving the wolf [...]."[17] In sum, there is a wide range of interpretations regarding the nature of estimative perception in Aquinas. While some interpreters consider it to come close to or to be some kind of conceptualisation and categorisation, others are more cautious, and some even reject such conclusions altogether.

What militates against the interpretation of the first group is that Aquinas himself is quite keen to stress the difference between the estimative power of nonhuman animals and the cogitative power of humans. Contrary to what

13 See Tellkamp (2006), 1351.

14 Pasnau (2002), 270.

15 Stump (1999), 379 and 391f. See also Stump (2003), 246f. Stump's argument is actually more sophisticated. In her opinion, there is a difference between *sensory cognition* (e.g. seeing something) and *perception* (i.e., seeing something *as* something). While humans have both, nonhuman animals lack the latter. This terminology is, in a sense, misleading, however, because 'perception' is usually employed for the kind of (sensory) cognition nonhuman animals have.

16 Pasnau (2002), 271. See also Perler (2004), 60, who says that the sheep perceives the wolf ("Das Schaf nimmt den Wolf zwar wahr") but does not grasp it *as* wolf ("es erfasst ihn nicht als Wolf") and is incapable of making general statements about the wolf ("ist nicht in der Lage, allgemeine Aussagen über den Wolf zu machen").

17 Pasnau (1997), 53f.

the reading of the first group seems to suggest, estimation is not the nonhuman *equivalent* of human cogitation. It is equal to it, of course, insofar as it enables nonhuman animals to perceive individual intentions just as the cogitative power makes humans perceive individual intentions. And in both cases, the perception of such intentions results in accidental cognition. Yet, what is different is that, according to Aquinas, estimation, in contrast to the cogitative power, is incapable of apprehending individuals 'as existing under a common nature' (*ut existens sub natura communi*). Therefore, the sheep 'cognises *this* lamb not insofar as it is this *lamb*' (*ovis cognoscit hunc agnum, non inquantum est hic agnus*) or *this* wolf not insofar as it is this *wolf*. Rather, it cognises a lamb 'insofar as it is to be nursed by it' (*inquantum est ab ea lactabilis*), an herb 'insofar as it is its food' (*inquantum est eius cibus*) and a wolf insofar as it is its enemy.[18]

One could argue that seeing food instead of herbs, enemies instead of wolves, and things that need to be nursed instead of lambs still involves conceptualising perceptual content in a way, even if the concepts which are used here are 'proto-concepts'. For although the (proto-)concept 'enemy' is broader or has a wider extension than the concept 'wolf', seeing an enemy is, nevertheless, a conceptualisation of perceptual content, and hence more than just seeing a grey, furry thing. Furthermore, one might say that it is by means of such (proto-)concepts that animals categorise the things they perceive, as they distinguish, for instance, between edible things, harmful things, and needy things.

Yet, what is problematic about this reading of Aquinas is, first, that it depends on attributing representations to nonhuman animals which are still more general than Aquinas himself seems to allow. He undoubtedly takes such animals to distinguish between different things and, if all goes well, not to mistake wolves for lambs. But he clearly says that they are incapable of grasping individuals 'as existing under a common nature', as has been stressed above. Accordingly, any representation that is superordinate to the representation of an individual thing seems to be denied to nonhuman animals. It does not matter that they have accidental cognition because if what is perceived accidentally is 'something universal' (*aliquid universale*), it is the intellect from which this

18 See Thomas Aquinas, *In Aristotelis librum De anima commentarium*, lib. II, lec. 13, n. 398, ed. Pirotta (1959), 101: "Differenter tamen circa hoc se habet cogitativa, et aestimativa. [...] Aestimativa autem non apprehendit aliquod individuum, secundum quod est sub natura communi, sed solum secundum quod est terminus aut principium alicuius actionis vel passionis; sicut ovis cognoscit hunc agnum, non inquantum est hic agnus, sed inquantum est ab ea lactabilis; et hanc herbam, inquantum est eius cibus."

cognition arises.[19] Therefore, it makes sense to save the term 'proto-concepts' for the intentions that are perceived by the human power of cogitation.[20] Second, in a passage of the *Summa theologiae*, Aquinas states that the sheep flees the wolf '*as if* it were an enemy by nature' (*quasi inimicum naturae*).[21] This can be read as pointing to the fact that the sheep does not conceptualise the grey furry thing it sees – neither as 'wolf' nor, more generally, as 'enemy'.

Aquinas' most explicit statement regarding how (or as what) nonhuman animals apprehend individual things is found, once more, in his commentary on *De anima*. In this text, he says that by virtue of estimation an individual thing is apprehended 'only insofar as it is the end or starting point of some action or passion' (*solum secundum quod est terminus aut principium alicuius actionis vel passionis*).[22] So the sheep perceives the wolf only insofar as it is the starting point of the emotion of fear and of the action of fleeing. Thus, the sheep's reaction to the wolf could be described by a simple stimulus-response model. According to this model, a certain reaction (e.g. flight) is triggered once a certain stimulus occurs (e.g. a wolf). Again, what is important to note is that in Aquinas' (and Avicenna's) theory such a stimulus is *not* a colour, shape, sound, or any other sensible form. It is not that the wolf, for example, looks or smells threatening or ugly to the sheep. Rather, the decisive stimulus is the intention of hostility that is perceived by the estimative power.[23] Surely, one could object that it is superfluous to posit such cognitive vehicles as intentions in addition to sensible forms. For why should the perception of forms be insufficient to trigger a reaction?[24] Yet, even if one rejects the theory of intentions, the stimulus-response model as such remains the same because a reaction is triggered, regardless of whether the stimulus is a sensible form or an insensible intention. Moreover, in both cases the stimulus is a particular, not a

19 See Thomas Aquinas, *In Aristotelis librum De anima commentarium*, lib. II, lec. 13, n. 396, ed. Pirotta (1959), 101.

20 This is the way in which, for instance, Davids (2017), 167f., employs this term.

21 Thomas Aquinas, *Summa theologiae* I, q. 78, a. 4, co., ed. Leonina V (1889), 256: "Sed necessarium est animali ut quaerat aliqua vel fugiat, non solum quia sunt convenientia vel non convenientia ad sentiendum, sed etiam propter aliquas alias commoditates et utilitates, sive nocumenta: sicut ovis videns lupum venientem fugit, non propter indecentiam coloris vel figurae, sed quasi inimicum naturae; et similiter avis colligit paleam, non quia delectet sensum, sed quia est utilis ad nidificandum. Necessarium est ergo animali quod percipiat huiusmodi intentiones, quas non percipit sensus exterior."

22 See n18 above.

23 See n21 above. This point is also stressed by Kenny (1993), 37, and Black (2000), 67.

24 This objection was raised, for instance, by John Duns Scotus, see Perler (2012b), 38f.

universal, and in both cases the reaction is a natural (or instinctual) reaction.[25] The sheep perceives the intention of hostility of *this* particular wolf and *this* intention triggers the reaction of flight.

Still, two questions can be raised here. The first concerns the instinctual nature of the sheep's reaction. Speaking of an instinctual reaction somehow implies that the reaction is innate or naturally inscribed in the sheep's behaviour. One might therefore get the impression that nonhuman animals come with a set of fixed reactions to all kinds of stimuli: sheep flee wolves but seek out other sheep and so forth.[26] But what about the example of the dog that somehow *learns* to fear sticks?[27] Obviously, dogs do not naturally dislike sticks. Rather, they begin to fear them once they have been beaten with one. In this case, the stimulus stays the same but the animal's reaction changes. Moreover, this kind of change seems to involve what is usually called generalisation, that is, the capacity to make inferences of the form 'If *this* stick is painful, *all* sticks are painful'.[28] This gives rise to the second question, namely, whether nonhuman animals are capable of generalising, as the example of the dog seems to imply.

To begin with the first question, neither Aquinas nor Avicenna have problems explaining how dogs come to fear something they have not feared before. While the sheep's reaction of flight when seeing a wolf arises out of 'natural caution' (*cautela naturalis*), the reaction of flight exhibited by a dog that has been beaten in the past when seeing a stick or a stone comes 'from experience' (*per experientiam*), as Avicenna puts it.[29] That is to say, while the sheep's estimative power always perceives the hostility of the wolf, even if it has never seen a wolf before, the dog's estimation learns to associate the intention of

25 This is why Aquinas also calls the estimative power 'natural estimation' (*aestimative naturalis*). See *In Aristotelis librum De anima commentarium*, lib. II, lec. 13, n. 397, ed. Pirotta (1959), 101.

26 Intentions are, of course, "relational properties," as Perler (2012b), 38, puts it. This explains why sheep flee wolves, whereas other wolves do not.

27 See Avicenna Latinus, *Liber de anima*, lib. IV, c. 3, ed. van Riet (1968), 39.

28 As mentioned before, Bennett (1989), 104, calls this a "one-many generalization."

29 See Avicenna Latinus, *Liber de anima*, lib. IV, c. 3, ed. van Riet (1968), 38f.: "Praeter hoc etiam animalia habent cautelas naturales. [...] Et per istas cautelas apprehendit aestimatio intentiones quae sunt commixtae cum sensibilibus de eo quod obest vel prodest; unde omnis ovis pavet lupum, etsi numquam viderit illum nec aliquid mali pertulerit ab illo [...]. Alius autem modus est sicut hoc quod fit per experientiam [...]; unde canes terrentur lapidibus et fustibus et similia." On this passage see also Piro (2005), 134f.; Tellkamp (2012b), 630; Kaukua (2014), 106f.; Juanola (2015), 350–353.

painfulness with sticks.[30] Its fear of sticks is nothing other than what is called a case of classical conditioning in modern psychology (the most famous example of which is Pavlov's dog).[31] In such a case, an animal is repeatedly confronted with a pair of stimuli. One of these stimuli is rather neutral at first, whereas the second stimulus is potent in that it arouses desire or aversion in the animal. So when Aquinas' and Avicenna's dog is repeatedly beaten with a stick it learns that the originally neutral stimulus 'stick' is now always accompanied by the stimulus 'pain'. Since pain is something the dog aims to avoid, it will begin to run away every time it perceives a stick.

Both Avicenna and Aquinas agree that the dog would not be able to establish a connection between sticks and pain if it lacked the faculty of memory (*memoria*).[32] The dog needs to realise that the stick it sees *now* caused it pain the *last time* it saw it. However, Aquinas is rather hesitant to claim that this memory-based association of stimuli amounts to what Avicenna (or his Latin translator, more precisely) calls 'experience' (*experimentum*). For Aquinas, experience is 'the gathering of many singulars' (*collatio plurium singularium*).[33] In this sense, it cannot be found in nonhuman animals because the mechanism of collation requires the cogitative power or 'particular reason' (*ratio particularis*). As was mentioned, nonhuman animals lack this power.[34] Nevertheless, they do 'participate in some experience but only to a very small degree' (*aliquid experimenti, licet parum, participare*), namely, insofar as they 'get accustomed' (*consuescunt*) to certain things. The dog thus gets used to the

30 One could also say that the sheep's reaction is based on 'inclinations that are innate, not learned' ("tendencias innatas, no aprendidas") as Juanola (2015), 350, puts it.

31 For an introduction see, for instance, Shettleworth (2010), 105–109.

32 See Avicenna Latinus, *Liber de anima*, lib. IV, c. 3, ed. van Riet (1968), 39, and Thomas Aquinas, *Quaestiones disputatae De veritate*, q. 24, a. 3, ad 7, ed. Leonina XXII.3.1 (1973), 687.

33 Thomas Aquinas, *In duodecim libros Metaphysicorum Aristotelis expositio*, lib. 1, lec. 1, n. 15, eds. Cathala & Spiazzi (1964), 8: "Supra memoriam autem in hominibus, ut infra dicetur, proximum est experimentum, quod quaedam animalia non participant nisi parum. Experimentum enim est ex collatione plurium singularium in memoria receptorum. Huiusmodi autem collatio est homini propria, et pertinet ad vim cogitativam, quae ratio particularis dicitur [...]. Et, quia ex multis sensibus et memoria animalia ad aliquid consuescunt prosequendum vel vitandum, inde est quod aliquid experimenti, licet parum, participare videntur." On the various meanings of '*experimentum*' in Aquinas see Barker (2012b).

34 Roling (2011), 237, slightly misreads Aquinas when he claims that Aquinas ascribes a *ratio particularis* to nonhuman animals.

fact that sticks cause pain and so runs away every time it perceives a stick.[35] Now since it does this every time why is this not experience proper? Apparently, the dog has learned that *this particular* stick and *that particular* stick cause pain. From these painful encounters with individual sticks it seems to have acquired the experience that sticks *in general* are painful. This is exactly what the second question is about: Is the dog capable of *generalising*?

As far as Avicenna's answer is concerned, there are at least two possible ways of interpreting his position.[36] They mainly differ with regard to the way in which an intention such as the harmfulness or painfulness of the stick is perceived by an animal. One way would be that the dog receives what might be called a 'stick-intention' every time it sees a stick. The cognition of intentions is thus understood as *passive reception*. Originally, this stick-intention is something like harmlessness: the dog is not afraid of sticks. Yet, after several painful encounters with sticks, the stick-intention changes from harmlessness to harmfulness. In this case, there is no generalisation involved because every time a dog sees a stick it perceives the particular intention of this particular stick, regardless of how the intention appears. Still, the cognitive process might also work in another way, namely, through the *active ascription* of intentions.[37] In this case, the dog does not perceive any intention from sticks at first, but only begins to ascribe the intention of harmfulness to sticks after having been beaten with one. This process involves generalisation since the dog, after the first painful encounter with a stick, ascribes harmfulness not only to the particular stick it has been beaten with, but to sticks in general.

Both readings have their strengths and weaknesses. The first reading which suggests that the cognition of intentions is a passive reception gains much plausibility from the fact that intentions are 'bound to the form of the sensible' (*ligata cum forma sensibili*) or 'mixed with sensibles' (*commixtae cum sensibilibus*), according to Avicenna.[38] Sensible forms thus piggyback on intentions. What is problematic about this reading is that it cannot explain a behavioural change without some more or less miraculous change in the intention. As long as the dog does not run away when it perceives a stick-form, the correspondent intention that it perceives seems to be harmlessness. This intention then changes

35 Albertus Magnus, *Quaestiones super De animalibus*, lib. VIII, q. 3, ed. Filthaut (1955), 189, also accounts for such behaviour by referring to 'the way of experience' (*via experientiae*). On 'experimental cognition' (*cognitio experimentalis*) in nonhuman animals see also his *De animalibus*, lib. XXI, tr. 1, c. 2, ed. Stadler (1920), 1327. For an overview of thirteenth-century accounts of experience in animals see Köhler (2014), 189–203.

36 I am grateful to Juhana Toivanen for helpful suggestions concerning these readings.

37 The active nature of the inner senses is stressed by Kaukua (2014), esp. 106–111.

38 See p. 53 n4.

to harmfulness when the dog has been beaten with a stick while the stick-form stays the same. How exactly this change in the intention comes about, though, remains rather obscure. Here the second reading seems to provide a more convincing explanation: the change in behaviour is based on a change in estimation. The estimative power ascribes a different intention to the stick-form because of a different experience with this form.[39] The problem with this reading is that the estimative power is usually said to *abstract* intentions rather than to *ascribe* them.[40] It abstracts them from the sensible forms with which they travel to the organs of the perceiver. Consequently, it seems incoherent to claim that the estimative power ascribes intentions to sensible forms.

In order to finally come to a coherent understanding of the process it is helpful to recall that Avicenna introduces the example of the dog and the stick as an instance of a particular 'mode' (*modus*) of estimative cognition, namely, cognition 'by experience' (*per experientiam*).[41] This mode should not be conflated with other modes of estimative cognition, such as the mode by which a sheep perceives a wolf as hostile. Of course, there is at least one thing these modes have in common: they are cases of incidental perception. The sheep perceives the wolf *as* hostile and the dog perceives the stick *as* harmful. What is different, though, is the genesis of this perception. The sheep does not need to learn that wolves are hostile (and, one might add, its chances of learning this are rather low anyway because its first encounter with a wolf is usually its last). The dog, by contrast, learns to associate the stimulus 'stick' and the stimulus 'pain'. What is crucial here is that both stimuli are sensible forms, not intentions. This is clear from both passages in the *De anima* in which Avicenna refers to the example of the dog. At the beginning of book IV, Chapter 1, he says that 'there is a power in animals in which the forms of sensible things are joined together'. One proof for the existence of such a power is the behaviour of dogs that are afraid of sticks. This they would not do, Avicenna argues, if 'the form of the stick would not be remembered in conjunction with the form of pain'.[42] Apparently, the perception of an intention plays no role here, at least not at this stage of the cognitive process.

39 This then is an instance of what Kaukua (2014), 106, calls "the subject-specificity of estimation." Estimation is subject-specific in that it triggers different reactions to like stimuli in different individuals and species: while some dogs flee from sticks, others do not; while sheep flee from wolves, wolves do not.

40 See D'Ancona (2008), 56f.

41 See p. 63 n27.

42 Avicenna, *Liber de anima*, lib. IV, c. 1, ed. van Riet (1968), 2: "Si autem non esset in animalibus virtus in qua coniungerentur formae sensatorum, difficilis esset eis vita, si olfactus non ostenderet saporem et si sonus non ostenderet saporem, et si forma baculi non rememoraret formam doloris ita quod fugiatur ab eo." It may seem strange to speak of the

The various steps of the cognitive process underlying the dog's behaviour are described in chapter 3 of book IV.[43] It begins, first, with the *perception of sensible forms in the external senses*, such as the visual perception of the stick-form and the perception of pain. What follows is, second, a *bundling of these forms in the common sense*. The form of the stick and the form of pain are now joined together. The third step is the *storing of this bundle of forms in the imagination*. Up to this point, no intention is involved in the process. This only becomes relevant in the fourth step which Avicenna describes as follows: 'An intention of a comparison of how they [i.e., the forms] are related to each other and a judgment about them are inscribed in memory'.[44] That is, the *inscription of the intention of those forms in memory* is where intentions enter the process. Memory is the power which 'by itself and naturally apprehends this', according to Avicenna. But what about estimation? Estimation is the faculty which, fifth and finally, brings about the *perception of the unity of forms and intentions*. It triggers the dog's reaction of flight as it perceives the harmfulness of the stick based on the conjunction of the stick-form and pain. It thus establishes a "causal chain between apprehension, passion, and behaviour," as Dominik Perler puts it.[45]

What still remains to be answered is the question of whether this process amounts to generalisation. As has been said, the dog seems to generalise insofar as it flees every stick after it has had a number of painful encounters with individual sticks. The process of learning or conditioning seems to have established the general rule 'All sticks are painful'. The problem with this description is that Avicenna does not refer to universal entities such as 'all sticks', 'every stick' or 'sticks in general' in his account of estimative perception. Rather, he stays at the level of particulars. Forms are particulars insofar as the form of

'form' of pain but what Avicenna has in mind here is the hurtful tactile sensation the stick produces in this situation. Interestingly, Albertus Magnus, *De homine*, eds. Anzulewicz & Söder (2008), 284, employs the term '*intentio*' in this context by saying that 'the intention of pain is joined together with the form of the stick' (*intentio doloris coniungitur cum figura baculi*). However, ibid., 269, reads: 'the dog apprehends the form of pain and flees' (*canis apprehendit formam doloris, et fugit*).

43 Ibid., c. 3, 39. My reconstruction of this process largely accords with Kaukua (2014), 107.

44 Ibid.: "(…) et descripta fuerti in memoria intentio comparationis quae est inter illas et iudicium de illa."

45 Perler (2012b), 37. Unfortunately, neither Perler nor Kaukau (2014), 106, who says that "an associated sensation of pain [is] *coupled with* a maʿnā [i.e., an intention; A.O.], such as 'harmful' or 'evil'," explain *where* exactly this intention comes from. Kaukua at least stresses the active nature of the estimation in this process although he does not explicitly suggest that the estimative power itself brings about the intention.

the stick that is perceived, for instance, by the sense of sight is the form of *this particular* stick. Not even humans can see sticks in general. Likewise, intentions are particulars because they are linked to matter, as mentioned before. Although intentions are abstractions to some degree, Avicenna emphasises that they are not abstracted 'from the accidents of matter' (*ab accidentibus materiae*) and so they cannot be anything but particular.[46]

Nevertheless, one could argue that it is incoherent to claim that the dog's reaction to the stimulus 'stick' involves no generalisation at all. Admittedly, the dog does not have to cognise general rules such as 'All sticks are painful'. At the same time, it could not learn to flee from sticks if it did not engage in what is called 'stimulus generalisation' in modern research. This means that it is capable of recognising a stimulus under slightly different conditions: it recognises one and the same stick despite a lighting change, for example, and it perceives something as 'stick' although the shape and colour of this particular stick differ slightly from that of another particular stick that it has perceived before.[47] Avicenna does not, of course, discuss this notion of generalisation. However, his account of estimation could accommodate it, because stimulus generalisation in the sense described above does not necessarily go beyond the level of particulars. What is required is that imagination, estimation, and memory somehow match the sensible form of the stick that is currently being received via the external senses with the sensible form of the stick that has been perceived in the past. So the dog runs away from a stick even though it has no cognitive access to the fact that all sticks are (or might be) painful.

Interestingly, Aquinas is more explicit than Avicenna as regards the question of generalisation at the sensory level. In a passage of the *Summa theologiae* he states that not only intellectual, but also sensitive powers can be 'carried to something universally' (*ferri in aliquid universaliter*).[48] For instance,

46 Avicenna, *Liber de anima*, lib. II, c. 2, ed. van Riet (1972), 119.

47 See, for instance, Stach (2003), 1, who remarks that "[m]ost animals live in an extremely rich sensory environment in which stimuli hardly ever occur twice in exactly the same way." Therefore, the recognition of a stimulus always requires some "perceptual abstraction," as Deeley (1971), 63, puts it.

48 Thomas Aquinas, *Summa theologiae* I-II, q. 29, a. 6, co., ed. Leonina VI (1891), 207f.: "Potest tamen aliqua potentia sensitiva, et apprehensiva et appetitiva, ferri in aliquid universaliter. Sicut dicimus quod obiectum visus est color secundum genus, non quia visus cognoscat colorem universalem; sed quia quod color sit cognoscibilis a visu, non convenit colori inquantum est hic color, sed inquantum est color simpliciter. Sic ergo odium etiam sensitivae partis, potest respicere aliquid in universali, quia ex natura communi aliquid adversatur animali, et non solum ex eo quod est particularis, sicut lupus ovi. Unde ovis odit lupum generaliter."

sight perceives not only one particular colour but *all* colours or colours *in general*. Similarly, an emotion like hatred 'can consider something in a universal way' (*potest respicere aliquid in universali*). Sheep hate wolves 'in general' (*generaliter*) and not only one particular wolf. The same could be said of dogs and sticks. For Aquinas, a thing is rejected 'because of its common nature' (*ex natura communi*) and not because it is this or that particular thing. Still, the question is whether he really thinks that a sheep flees a wolf because it realises that the individual it sees shares in the common nature of wolves. Of course, the intention of hostility is found in all wolves and thus, so to speak, carries information about the common nature of wolves.[49] Nevertheless, the intention itself is an *individual* object, according to both Avicenna and Aquinas. Therefore, perceiving the wolf's intention of hostility is different from cognising that hostility is found in all wolves, and it is even more different from abstracting the universals 'hostility' or 'wolf'. In short, *cognising universally* is to be distinguished from *cognising universals*.

To sum up, Aquinas does, in fact, consider many cases in which animals grasp features that are common to several particulars. But he makes it quite clear that they do not cognise such features as being *common* or *universal* to several individuals, but rather as being features of this or that *particular* thing. Therefore, it seems rather doubtful that he ascribes the capacities of conceptualisation, categorisation, and generalisation to nonhuman animals. One could argue that he does, if one applies those terms in a relatively loose sense. But then the question arises: is this really helpful in order to clarify Aquinas' position? What seems more helpful is to compare his views to those of some other authors from the thirteenth and fourteenth centuries who also dealt with various questions concerning universal cognition and concept formation in nonhuman animals.

49 This is also stressed by Rubini (2015), 487, who claims that intentions are 'universal contents' ("allgemeine Inhalte").

CHAPTER 9

Intentions and Quiddities (Albertus Magnus)

Albertus Magnus is a good starting point not so much because there is a close biographical connection between him and Aquinas but mainly because their positions overlap in some respects, while they diverge in others. Like Aquinas, Albert considers the estimative power to bring about accidental cognition in nonhuman animals.[1] So when the sheep sees a wolf, it receives sensory information not only via its external senses but also via its internal senses, most importantly through estimation. It is this power that provides cognitive access to features like hostility, that is, something which is common to all wolves (at least from the perspective of the sheep), hence different from the particular shade of grey of this or that particular wolf. It has already been mentioned that the universality of a feature like hostility easily gives rise to the impression that the estimative power grasps something universal. Albert shares this impression and one of the arguments he gives in his *De homine* is that what determines the sheep's reaction of flight 'is acquired from every wolf' (*accipitur ab omni lupo*).[2] Similarly, the sheep is unafraid of every shepherd, not only of a particular one. Therefore, it seems to grasp a universal, something that is found in more than just one individual.

Although this is a plausible claim Albert ultimately denies that nonhuman animals grasp universals by virtue of their estimative power. It is true, he says, that a sheep flees every wolf but this behaviour is based on an instinctual process rather than on rational 'enquiry and planning' (*inquisitio et consilium*).[3] While humans usually have rational control over their reactions and show

1 On Albert's theory of inner senses see Steneck (1974); Black (2000), esp. 63–66; Anzulewicz (2002); Di Martino (2008), esp. 69–84; Tellkamp (2012a).

2 Albertus Magnus, *De homine*, eds. Anzulewicz & Söder (2008), 294f.: "Item, quicquid uno modo ens accipitur ab omni sensibili eiusdem speciei, hoc videtur esse universale, sicut homo in una ratione accipitur a Socrate et Platone; id autem quod determinat fugiendum esse in ove secundum unum modum accipitur ab omni lupo, et non fugiendum esse secundum unum modum ab omni pastore; ergo hoc erit universale."

3 Ibid., 295: "Ad aliud dicendum quod aestimativa magis imitatur naturam in movendo quam scientiam et artem intellectus practici. Unde licet uno modo ovis accipiat inimicum in omni lupo, tamen non utitur inquisitione et consilio in modo fugiendi quemadmodum homo; sed omnis ovis fugit uno modo omnem lupum. Et hoc est quod dicit Damascenus quod 'bruta potius aguntur a natura quam agant'. Et idem dicit Avicenna quod instinctu naturae moventur bruta et non ad rationem apprehensi, et ideo rationem universalis non considerant."

 | DOI 10.1163/9789004363779_013

some flexibility in how they respond to a given stimulus, nonhuman animals lack this kind of control and always react to the same stimulus 'in the same way' (*uno modo*). For Albert, this behavioural uniformity provides evidence for the fact that nonhuman animals do not actively acquire universals from the perception of several individuals. Yet, even if one agrees with Albert in this regard, one could still ask him to further describe this perception of individuals. What does the sheep perceive? A wolf, an enemy, or a grey furry thing?

For Aquinas, as has been said, the estimative power makes the sheep see the wolf as the starting point of the emotion of fear and the action of flight.[4] Whether this amounts to more than to seeing a grey furry thing, but to less than seeing a wolf remains a matter of debate. Albert, by contrast, clearly appears to teach that the sheep does not simply see a grey furry thing, but a wolf or an enemy. Just as humans see that this black thing over there is a human being or a friend, 'a sheep sees that this hairy brownish thing is a wolf and that that hairy thing is a dog and [by its estimation] it judges one to be friendly and the other one to be hostile', Albert writes.[5] Similarly, a wolf perceives this or that individual to be its offspring just as humans perceive this or that person to be the son of Dion, as he puts it in his commentary on *De anima*.[6]

It thus seems as if, in Albert's opinion, nonhuman animals are capable of conceptualising the contents of their perceptions, at least insofar as they do not simply see coloured patterns, such as hairy brownish things, but objects, such as wolves or cubs. Their estimative power reaches well beyond the surface of sensible forms and expands into 'the proper essence' (*propria essentia*) of things.[7] It does not simply abstract sensible forms as the external senses, but 'substantial forms' (*substantiales formae*), as Albert states.[8] Therefore, one

---

4 See p. 61 n18.

5 Albertus Magnus, *De anima*, lib. 2, tr. 3, c. 5, ed. Stroick (1968), 104: "Sed secundum accidens sensibile est, quod propriam essentiam in sensum non imprimit nec primo nec secundo, sed cum sensibili proprio et communi accipitur, et non sine aestimatione vel cogitatione, sicut sunt intentiones, de quibus tertio gradu abstractionis diximus, sicut quod hoc album est Diarii filius vel hoc nigrum est amicus vel homo vel animal, sicut videt ovis, quod hoc pilosum fulvum est lupus et hoc pilosum est canis, et unum aestimat amicum et alterum inimicum."

6 See Albertus Magnus, *De anima*, lib. 3, tr. 1, c. 2, ed. Stroick (1968), 167: "[…] numquam contingit cognoscere, quod iste est filius Dionis, nisi notitia habeatur filiationis, secundum quod est in isto, nec umquam lupus miseretur nato suo, nisi habeat cognitionem et huius individui et quod hoc individuum est natus eius."

7 See n5 above.

8 Albertus Magnus, *De anima*, lib. 2, tr. 3, c. 4, ed. Stroick (1968), 101f.: "Tertius autem gradus apprehensionis est, quo accipimus non tantum sensibilia, sed etiam quasdam intentiones quae non imprimuntur sensibus […]. Et tale est, quod accipimus hunc esse filium Deonis et esse

might get the impression that, in his view, nonhuman animals' perception of the world is not very much different from our own, because, despite their lack of intellectual powers, they do seem to gain cognitive access to the very essences of things by way of perceiving intentions, just as we do when grasping the quiddity of something.[9]

However, one needs to be careful here. Albert draws a clear distinction between the abstraction of substantial forms and the abstraction of quiddities. First, there is a difference with regard to the mode of apprehension. Although intentions, unlike sensible forms, 'are not imprinted into the senses' (*non imprimuntur sensibus*), their perception is, nevertheless, 'mediated by the perception of sensibles' (*mediantibus sensibilibus*).[10] Put differently, substantial forms are always apprehended in combination with material sensibles. Quiddities, by contrast, 'are denudated from all appendices of matter' (*denudatas ab omnibus appendiciis materiae*).[11] They are thus entirely immaterial. Consequently, there is, second, a difference with regard to the cognitive powers that apprehend them. While substantial forms are perceived by a material power, such as estimation, the apprehension of quiddities is left to the immaterial intellect alone.

Because of this restriction, even relatively highly-developed nonhuman animals are incapable of grasping the quiddities of things. The most prominent examples are pygmies with whom Albert deals at length in his commentary on *De animalibus*.[12] In his opinion, pygmies show a surprisingly high degree of perfection and come very close to humans, both with regard to their physique and with regard to their cognitive capacities. They do, for instance, gather many things in their memories, are capable of learning, and also possess some sort of 'language' (*loquela*), which is why 'they seem to have something imitating reason' (*videtur aliquid habere imitans rationem*).[13] Nonetheless, they still stand at

---

agnum vel hominem, aliud autem esse lupum vel leonem, secundum quod substantiales formae mediantibus sensibilibus et non separatae ab ipsis apprehenduntur."

9 This is actually what Sobol (1993), 117, seems to suggest.

10 See Albertus Magnus, *De anima*, lib. 2, tr. 3, c. 5, ed. Stroick (1968), 104.

11 Ibid.: "Quartus autem et ultimus gradus est, qui apprehendit rerum quiditates denudatas ab omnibus appendiciis materiae nec accipit ipsas cum sensibilium intentionibus, sed potius simplices et separatas ab eis. Et ista apprehensio solius est intellectus, sicut est intellectus hominis [...]."

12 On pygmies in medieval philosophy in general and in Albert in particular see Koch (1931); Janson (1952), esp. 84–89; Köhler (1992) and (2008), 419–443; Thijssen (1995); Friedrich (2009), 138–141; Roling (2010), 486–498; Salisbury (2010), 122–124.

13 Albertus Magnus, *De animalibus*, lib. XXI, tr. 1, c. 2, ed. Stadler (1920), 1328: "Quaedam autem in tantum vigent in disciplina auditus quod etiam sibi mutuo suas intentiones

least one step below humans on the ladder of beings because they lack the faculty of reason. Therefore, their cognitive capacities originate mainly from what Albert calls 'a shadow of reason' (*umbra rationis*).[14] This power is superior to the estimative power of other nonhuman animals, but, nevertheless, according to Albert, inferior to the human intellect.[15] Its inferiority is most evident in regards to universal cognition because, in contrast to humans, pygmies are incapable of 'eliciting universals' (*elicere universalia*). Thus, their use of signs, both spoken and heard, is always bound to immediate sensible impressions. Unlike human beings, the pygmy 'neither argues nor talks about the universals of things but rather its voices are directed at the particular things about which it talks' (*non disputat nec loquitur de universalibus rerum, sed potius suae voces diriguntur ad res particulares de quibus loquitur*).[16]

Admittedly, it is difficult to understand how this 'language' of pygmies really works. But there are at least two ways in which Albert's description can be understood. One way is to say that the pygmy language works without universals or universal concepts at all. In this case, every sign refers to nothing other than one particular thing. So if a pygmy says, for instance, 'baobab', another pygmy will look at or go to the particular tree which they habitually call 'baobab'. This pygmy will not, however, go to or look at another tree of the same species because the word 'baobab' refers to one particular tree only. Thus, according to this reading, the names (or signs) pygmies employ are not universal in meaning. However, one might also read Albert as saying that pygmies do, in fact, talk

---

significant, sicut pigmeus qui loquitur, cum tamen sit irrationabile animal: et ideo quantum ad animales virtutes, post hominem videtur pigmeus esse perfectius animal: et videtur quod inter omnia animalia plus confert memorias suas et plus percipit de signis auditus, ita quod videtur aliquid habere imitans rationem, sed ratione caret." The pygmies' capacity to speak is also mentioned in *De animalibus*, lib. I, tr. 1, c. 3, §44, ed. Stadler (1916), 17, and lib. VII, tr. 1, c. 6, §62, 521f.; *Liber de natura et origine animae*, tr. 1, c. 1, ed. Geyer (1955), 1. On Albert's account of the language of pygmies (and of other animals) see especially Resnick & Kitchell (1996), 57–61; Tellkamp (2013), 214–217; Köhler (2014), 487f.; Toivanen (forthcoming).

14 On the origin and meaning of this idea see Köhler (2008), 434f., and (2014), 199. See also Roling (2011), esp. 231–233.

15 See Albertus Magnus, *De animalibus*, lib. XXI, tr. 1, c. 2, ed. Stadler (1920), 1329.

16 Ibid., 1328: "Ratio enim est vis animae discurrendo per experta ex memoriis accepta, per habitudinem localem aut sillogisticam, universale eliciens et ex illo principia artium et scientiarum per similes habitudines conferens: hoc autem non facit pigmeus, sed ea quae accipit cum auditu, non separat a sensibilium intentionibus [...]. Et ideo pigmeus licet loquatur, tamen non disputat nec loquitur de universalibus rerum, sed potius suae voces diriguntur ad res particulares de quibus loquitur. Causatur enim loquela sua ex umbra resultante in occasu rationis." See also Köhler (2008), 430–438.

about many things other than universals. In this case, they are capable of applying the name 'baobab' *universally*, that is, to more than one particular tree, but fail to talk about 'baobab' *as a universal*. So they fail to engage in *metalanguage*, as one could put it, since they are incapable of grasping 'the quiddities of things' (*quiditates rerum*).[17]

What speaks in favour of the first reading is Albert's statement that, unlike human language, the language of pygmies originates 'from the simple imagination and not from the intellect' (*ex ymaginatione simplici et non ex intellectu*).[18] Therefore, the cognitive vehicles of their language, so to speak, are particulars because the imagination is a material power, which entails that it is incapable of working with universals. What militates against the first reading, however, is that if pygmies really employed a particular name for each particular thing, their language would be rather impractical. They would need a plethora of words to name things, as no name could be applied to more than one thing. Moreover, it seems to be unlikely that they are, on the one hand, capable of distinguishing things of the same type from things of another type, but, on the other hand, incapable of employing the same name for various individuals of the same type.

For although in *De animalibus* Albert does not explicitly say that pygmies (and other nonhuman animals) are capable of discerning things of one type from things of another type, he does make a brief remark on animals' ability to discern in his *Quaestiones super De animalibus*. In this text, he explains that discerning (*discernere*) is actually 'a work of reason' (*opus rationis*).[19] Yet, he concedes that those animals that lack reason nevertheless engage in discerning things, since camels and horses, for instance, obviously distinguish their offspring or parents from other members of their species. Similarly, birds discern between stones and grains, and sheep distinguish wolves from lambs or human beings.[20] Their capacity to discern things is, of course, derived from their estimative power. Therefore, it is 'a discernment between *individual* intentions' (*discretio inter intentiones individuales*) because the intentions they perceive are particulars, not universals. However, generality is involved insofar as their estimation enables them 'to discern between singular things *universally*' (*discernere inter singularia universaliter*). Put differently, the perception of,

---

17 Ibid.: "Et ideo pigmeus nichil omnino percipit de rerum quiditatibus nec umquam percipit habitudines argumentorum: et sua locutio et [*corr.* est; A.O.] sicut locutio morionum qui naturaliter stulti sunt eo quod non perceptibiles sunt rationum."

18 *De animalibus*, lib. IV, tr. 2, c. 2, ed. Stadler (1916), 400.

19 Albertus Magnus, *Quaestiones super De animalibus*, lib. VIII, q. 39, ed. Filthaut (1955), 200.

20 See ibid., 201.

say, a 'stone-intention' makes a bird discern stones from grains, because when seeing the latter it perceives a 'grain-intention' and this happens in a universal way insofar as stones in general are distinguished from grains in general. Again, what is important to note is that this is something different than distinguishing the universal 'stone' from the universal 'grain'. And while most animals are capable of *distinguishing universally*, only humans are capable of *distinguishing universals*.

Nonetheless, the perception of individual intentions by estimation might form not only the basis of nonhuman animals' capacity of discerning in general, but also of the language of pygmies in particular. This is because, although pygmies clearly lack cognitive access to universals, they might still possess the ability of applying names in a universal way. This could work as follows: every time a pygmy sees, for instance, a baobab tree, it perceives a 'baobab tree-intention'. The perception of this intention is linked to the utterance 'baobab' and this link works in both directions so that when seeing a baobab tree the pygmy knows that it can inform other pygmies about this tree by saying 'baobab'. Likewise, when it hears another pygmy say the word 'baobab' it knows what the other pygmy is talking about because the utterance 'baobab' is linked to the baobab tree-intention and thus produces the representation of a baobab tree in its imagination.[21] So, in a way, the pygmy language is based on a simple stimulus-response reaction: either the stimulus is an intention that triggers a linguistic response or the stimulus itself is a linguistic expression which then triggers a certain representation. This might be the reason why Albert considers the language of pygmies to be based on 'an instinct of nature' (*instinctus naturae*), in contrast to human language, which is based on reason.[22]

Thus, it seems that Albert, like Aquinas, does not ascribe universal cognition in terms of *cognising universals* to nonhuman animals, not even to such highly developed nonhuman animals as pygmies. However, it needs to be noted that he grants them a certain kind of 'experimental cognition' (*cognitio experimentalis*). This might seem surprising at first because experience (*experimentum*) is usually the cognition of what is common to many things. In other words, one is experienced if one has acquired some knowledge about certain universal aspects of things. For instance, if a physician realises that aspirin reduces the fever of many patients, he comes to know that aspirin is a fever-reducing

21 This, of course, requires both a storage of intentions as well as a storage of representations or images of things but Albert has no doubt that pygmies and most other animals possess the faculties of imagination and memory.

22 See Albertus Magnus, *De animalibus*, lib. 1, tr. 1, c. 3, ed. Stadler (1916), 18. See also Roling (2010), 495.

medication, not only in one particular case but in many, if not in all cases. According to Albert, this kind of knowledge can also be found in nonhuman animals. In particular, he mentions the example of a weasel that has been bitten by a snake and treats the bite with an endive leave.[23] The weasel thus seems to possess some kind of universal knowledge about the healing power of endive leaves. The question now is whether this experimental cognition amounts to the cognition of a universal.

Albert's reply is pretty clear: although the weasel does participate in experience to some extent the experimental cognition it has is only 'an experimental cognition in singulars' (*experimentalis cognitio in singularibus*). The experience of nonhuman animals is thus different from the full-fledged experience of human beings because the latter involves universals, while the former does not.[24] But is this a plausible distinction? Why is the weasel's knowledge about the healing power of endive leaves less universal than the physician's knowledge about aspirin? Albert does not explicitly address this question, but his answer might be something like the following: even though the weasel possesses universal knowledge about endive leaves in some sense, it does not acquire the universal 'endive'. Rather, it simply remembers that last time it had been bitten by a snake the bite was healed by the treatment with an endive leave. Surely, this supposes that the weasel recognises several elements in this situation: the snake, the bite, the endive leaf, etc. More precisely, it must be capable of recognising certain similarities between those elements. Otherwise it could not know that the bite it has now is the same kind of bite from which it suffered some time ago and so forth.

Nevertheless, this process of recognition does not require (or lead to) the cognition of universals because the weasel simply cognises that *this particular* bite is like *that particular* bite and that *this particular* leaf is like *that particular* leaf. The physician's knowledge, by contrast, amounts to something like the universal proposition 'Aspirin always reduces fever'. So even though the effect

---

23 Albertus Magnus, *De animalibus*, lib. XXI, tr. 1, c. 2, ed. Stadler (1920), 1327: "Quaedam autem animalium videntur aliquid licet parum experimenti participare. Experimentum namque ex multis nascitur memoriis quia eiusdem rei multae memoriae faciunt potentiam et facultatem experimenti: et nos videmus quod multa animalia praeter hominem aliquid experimentalis habent cognitionis in singularibus, sicut quod mustela pugnans cum serpente vulnerata contra venenum accipit folium endiviae quae a quibusdam rostrum porcinum vocatur: et alia talia multa induximus in praecedentibus quae faciunt animalia." Similar stories about birds are told in ibid., lib. VIII, tr. 2, ch. 2, ed. Stadler (1916), 590; *Metaphysica*, lib. I, tr. 1, c. 6, ed. Geyer (1960), 10.

24 See ibid.: "Sufficienter autem non participant experimento quia non veniunt per experimentum ad universale et artem et rationem, sed tamen secundum aliquid participant experimento, ut iam diximus."

is the same – both the weasel and the physician know how to medicate a certain illness – the ways in which they (mentally) represent what they do differ: the physician's knowledge involves universals, while the weasel's does not. In sum, Albert also seems to stick to the view that the boundary between the cognition of humans and nonhuman animals is largely identical with the boundary between universals and particulars. However, this does not mean that their behaviour does not often resemble or come close to human behaviour.

CHAPTER 10

Elevated Intentions and Common Forms (Pseudo-Peter of Spain)

Both Aquinas and Albertus Magnus follow Avicenna in making estimation one of the most important internal senses in nonhuman animals, as we have seen. In their view, estimation brings about certain accidental cognitions, triggers reactions, and is involved in distinguishing individuals. This happens universally or generally insofar as, for instance, sheep flee wolves in general and generally discern them from fellow sheep. However, they do not possess the concept or cognise the universal 'wolf'. For although the perception of intentions is an abstractive process, it is less abstractive or less complex an abstraction than the abstraction of universals by virtue of intellect and reason. Therefore, nonhuman animals engage in distinguishing between things of different types, yet they do so without actually cognising that there is something like a type. If they show the same reactions to the same stimuli, it happens because an individual stimulus, such as the intention of hostility, always triggers the same reaction, such as flight. It does not happen, however, because an individual wolf is identified as belonging to a general type, such as the species of wolves.

As mentioned before, there are, of course, a number of objections which can be raised against such a theory. Some objections were indeed raised (at least implicitly) in a thirteenth-century commentary on Aristotle's *De animalibus*. Its author is unknown, but is usually referred to as Pseudo-Peter of Spain.[1] Briefly speaking, Pseudo-Peter doubts that nonhuman animals are incapable of grasping the species (or kind) an individual belongs to. In fact, he thinks that they cognise both the species (or general substance) as well as the individual substance of a thing. One of the arguments he presents is that a sheep responds to the stimulus *wolf* rather than to the stimulus *this wolf*. That is, its behaviour resembles the behaviour of a physician who gives the same medicine to every patient suffering from the same illness. Of course, the patients are individuals and hence differ in various respects. One might be small and brown-haired, while another is tall and black-haired. Yet, what matters is that

1 Köhler (2014), 361, dates the commentary to before 1260. I am very grateful to him for providing me with the transcriptions of several of the questions of this commentary. On the different manuscripts see Köhler (2008), 25, n. 64. For a summary of Pseudo-Peter's position see Köhler (2014), 361–364.

 | DOI 10.1163/9789004363779_014

they are individuals *of the same kind* insofar as all of them suffer from the same disease.[2] Similarly, it does not matter for the sheep whether individual wolves differ slightly with regard to certain accidents. For what it knows is 'the common form' (*forma communis*) of wolf and 'through one likeness of wolf it cognises all wolves' (*per unam similitudinem lupi cognoscit omnes lupos*).[3] In the opinion of Pseudo-Peter of Spain, it is thus wrong to claim that a sheep flees *this individual* wolf. For even though it does, of course, flee an individual, it flees not because the wolf it perceives is *this individual* wolf. Rather, it flees because the individual is an individual *of a certain kind*, namely, a member of the *species* of wolves. Since the sheep responds to the general substance of the stimulus rather than to its individual substance, it engages in a 'universal grasp' (*apprehensio universalis*), according to Pseudo-Peter of Spain.[4]

Yet, he argues, animals also attend to the 'discrete substance of a thing' (*discreta substantia rei*). As he points out, they know to differentiate between individuals of the same kind. There are cases in which they do not show similar responses when confronted with similar stimuli. For instance, a lion usually responds with aggressive behaviour to the stimulus 'human being', but there is at least one human being towards whom it is obedient, namely, its tamer. For Pseudo-Peter, this is evidence for the fact that the lion, on the one hand, perceives its tamer as a member of the species of humans but, on the other hand, it cognises him as an individual person, too. This is because its cognition goes beyond the mere perception of accidents but can also attend to accidental features.[5] However, Pseudo-Peter is well aware of the fact that this is incompatible with the usual distinction between sensory and intellectual cognition. While

2 See Pseudo-Peter of Spain, *Commentum super libros de animalibus* (Venetian redaction), fol. 151va: "Sicut dictum est superius, animalibus brutis debetur mutuus aspectus cum rebus aliis, ut ab aliis recipiant iuvamenta et nocumenta. Ista non solum currunt a parte complexionis, sed circa substantiam et speciem. Unde ratione speciei ovis discernit lupum, et sic de aliis. [...]. Hoc idem est in medicinis, quia operantur a specie."

3 Ibid., fol. 152rb: "Preterea omnis apprehensio intentionis, que una existens ad multa nata est applicari, est intentio universalis vel per modum universalis. [...] Sed intentiones, per quas bruta cognoscunt animalia, sunt huiusmodi, ut agna per unam similitudinem lupi cognoscit omnes lupos." That the sheep 'compares the common form to the particular' (*comparat formam communem ad particulare*) is mentioned on fol. 153ra.

4 See ibid., 152ra-b: "Omnis acceptio speciei rei non apropriate individue substantie rei est acceptio universalis. [...] Sed hec est acceptio lupi ab agna, quia lupus, non quia ille lupus."

5 See ibid., 151vb: "Omnis substantia cognitiva, que circumscriptis accidentibus ratione speciei sequitur rem apprehensam et fugit et obedit vel inobediens est ei, apprehendit discretam substantiam rei et non solum accidentia. [...] Hoc patet, quia anima agne fugit lupum ratione lupi, non fugit canem, ymo ratione speciei sequitur sociam; ymo ovis obedit homini quia homo, et leo obedit magistro ratione persone individue. Ergo fit discretio ratione speciei et circa speciem. H<ec> supra accidentia." See also fol. 152vb: "Sed bruta discernunt rem a re

the latter is usually taken to consist in grasping substances and universal aspects, such as the species of an individual thing, the former 'cognises accidents and not the substance' (*cognoscit accidentia et non substantiam*). And since nonhuman animals lack intellectual powers, they also lack the cognition of species and substances.[6]

Nonetheless, he holds on to the view 'that brute animals cognise the species and substances of things' (*quod bruta cognoscunt species rerum et substantias*). In his opinion, this is a consistent view to hold if one takes into account that, as according to Averroes, there are two kinds of cognition of substances: an intellectual one by which universal intentions – 'the universal quiddities of things' – are grasped and a sensory one by which individual intentions are grasped.[7] So, in his view, it is possible to cognise the substance of a thing by way of perceiving an individual intention, but, in this case, the substance is grasped as connected to an individual thing. Still, the question is why this qualifies as 'universal grasp' (*acceptio universalis*) because, first, what is cognised is an *individual* not a *universal* intention. Second, if the intention is grasped as connected to an individual thing, the grasp is *material* and, unlike the universal grasp of something, does not go 'beyond matter' (*preter materiam*).[8]

To this question Pseudo-Peter replies as follows. First, he argues that the intentions received by the senses are not simply individual intentions but 'elevated individual intentions' (*intentiones individuales elevatas*), as he puts it. Unfortunately, he does not further explain how those *elevated* intentions compare to mere individual intentions, on the one hand, and to universal

---

sine accidentibus, ut hominem ab asino, et discernunt hunc hominem ab illo sub accidentibus. Ergo sic discernunt bruta."

6 Ibid., 151vb–152ra: "Duplex est cognitio. Quedam est rei per principia vera eius; hec solum debetur intellective anime; per hoc probat intellectum inesse homini. Unde cognoscit speciem non secundum quod stans vel huiusmodi. Alia est cognitio per accidentia; hec est sensitive, ut dicunt [Avicenna et Algazali; A.O.], quod cognoscit aliquem in quantum sedet vel huiusmodi. Sed cognitio bruti sensitiva est. Ergo solum cognoscit accidentia et non substantiam."

7 Ibid., 152ra: "Solvit Averroes: Cognitio substantie est duplex. Una est, que cadit super quidditates rerum universales, que intentiones universales ab omnibus individuis separate sunt et ab omnibus dependitiis accidentium. Hec intellective. Alia est cognitio, que cadit super intentiones individuales; hec inest brutis et multis, et hec est sensibilis virtutis, ut ipse dicit."

8 See ibid., 152vb–153ra: "Sed contra. Averroes dicit: Virtus cognitiva duplex. Una distinguit intentiones universales; hec intellectus. Alia est, que distinguit intentiones individuales; hec sensitiva. Hec competit bruto. Sed nulla virtus, que non habet acceptionem universalis, discernit ea. Sensitiva huiusmodi, ut dicit Averroes. Hoc di<ci>t Avicenna et Algazel in multis locis. [...] Sed acceptio intentionis individuorum est acceptio materialis prout res in materia corporali. Acceptio universalis in sua puritate est acceptio preter materiam. Hec fuit virtutis materialis, illa intellectus."

intentions, on the other hand. All he says is that they 'differ in degree' (*distant per gradus*) from those intentions that are received by the intellect. Thus, they 'do not reach the level of elevation as those which are in the intellect' (*non attingunt elevationem, que apud intellectum*).[9] Second, he argues that one should make a distinction between two kinds of universal cognition. One kind is purely intellectual and totally 'abstracts from matter and material principles' (*abstrahit a materia et principiis materialibus*). The other kind 'does not rise above material things' (*non erigitur ultra principia materialia*). Still, it is by this kind of universal cognition that nonhuman animals have access to what he calls 'the common form' (*forma communis*) of a thing, such as the common form of lamb or wolf. Moreover, they are capable of comparing such common forms to the particular thing they see, namely, *this* lamb or *this* wolf.[10] Thus, their cognition is clearly distinct from intellectual cognition but, nevertheless, universal in some respect.

In sum, the position of Pseudo-Peter of Spain does not differ from the views of Aquinas and Albert insofar as he also denies the cognition of universals in terms of universal quiddities to nonhuman animals. Yet, unlike Aquinas and Albert, he does not hold that universal cognition in such animals is restricted to cognising universally in the sense explained above. Rather, he emphasises that they engage in some sort of '*acceptio universalis*' of individuals which is, however, not as abstractive as the cognition of quiddities. The individual intentions that are cognised by nonhuman animals only differ in degree from those that are grasped by the intellect, he says, and they are 'elevated' in comparison to purely individual intentions. Yet, he also sticks to the view that there is a distinction between the sensory and the intellectual realm. And even though he seems sympathetic to saying that the distinction between those realms is not entirely equal to the distinction between particularity and universality, he still thinks that there is a difference between the cognition of universals that is found in humans and the kind of universal cognition one finds in nonhuman animals.

---

9 See ibid., fol. 153ra: "Dicendum, quod virtus sensitiva non accipit solum intentiones individuales, sed intentiones individuales elevatas. Iste tamen intentiones, quas accipit, non attingunt elevationem, que apud intellectum, sed distant per gradus."

10 Ibid.: "Acceptio intentionis universalis duplex. Una, que abstrahit a materia et a principiis materialibus; hec debetur anime intellective [...]. Alia est, que non erigitur ultra principia materialia. Unde si offeratur ei sub principiis spiritualibus res, non cognoscet . Et ista coniungit duo individualiter accipiendo, ut animam bovis cum corpore bovis. Alia est prout comparat formam communem ad particulare, ut agnam ad hanc agnam; hoc non fit sine materia, et hec est anime sensitive."

CHAPTER 11

Vague Particulars as Universals (Roger Bacon)

The position of Pseudo-Peter of Spain might strike one as comparatively original, if not even radical, in the sense that it ascribes a rather high capacity for the perception of what he calls 'elevated intentions' to animals. Yet, if one compares his statements on universal cognition in nonhuman animals to the statements of Roger Bacon, it seems as if Bacon goes even farther. At least this is the impression one can get from Bacon's treatise on optics, which is better known as *Perspectiva*. One passage in particular has attracted the attention of modern interpreters concerned with this topic:

> But it is certain that a dog recognises a man it has seen before when it sees him again and monkeys and many other nonhuman animals do this. And they distinguish things they have seen before and of which they have memory, and they cognise one universal from another, such as man from dog or from wood. And they distinguish individuals of the same species.[1]

Neither the beginning nor the end of this passage is particularly startling. All Bacon says is that dogs, as well as other nonhuman animals, are capable of making certain distinctions. First, they are able to distinguish between known and unknown individuals, as it were. For instance, they distinguish between a man they have met before, say, their owner, and a complete stranger. Second, they also manage to differentiate individuals whom they know equally well, such as their owner and his wife. However, in the central part of this passage Bacon adds that 'they cognise one universal from another, such as man from dog or from wood'. Although the wording of this passage is relatively clear, the crucial question now is how this is to be understood. Does Bacon really want to say that nonhuman animals do not only cognise particulars, such as *this* piece of wood or *this* man, but universals, such as 'wood' or 'man', too?

One possible answer is to say that he does. If this is correct, Bacon would indeed hold a very radical position because his view then amounts to saying that the cognition of universals is not dependent on the possession of intellect and reason which is why it can be found in those animals that lack those powers,

1 Roger Bacon, *Perspectiva*, pars II, dist. 3, c. 9, ed. Lindberg (1996), 246: "Sed constat canem cognoscere hominem prius visum cum iterum viderit eum, et simia et bestie multe sic faciunt. Et distinguunt inter res visas quarum habent memoriam, et cognoscunt unum universale ab alio, ut hominem a cane vel a ligno, et individua eiusdem speciei distinguunt."

 | DOI 10.1163/9789004363779_015

too. By and large, this is the answer which many of Bacon's modern interpreters seem to give. For instance, Theodor W. Köhler claims that Bacon ascribes 'a form of universal cognition' ("eine Form von Allgemeinerkenntnis"[2]) to nonhuman animals. Similarly, both Peter Sobol and Bernd Roling take Bacon to say that dogs and other animals have cognitive access to universals, such as 'man'.[3] If read in this way, Bacon's statement is, indeed, "an unexpected remark," as Jeremiah Hackett puts it,[4] because it thwarts the expectation of many readers according to which the cognition of universals is an intellectual operation. But maybe that is exactly the point Bacon wants to make. On the one hand, he clearly denies intellect and reason to nonhuman animals.[5] On the other hand, – and in contrast to most medieval authors – he grants them a cogitative power (*vis cogitativa*). It is this very faculty by means of which they 'cognise one universal from another', in his view.[6] Now since the cogitative power is responsible for 'a certain collecting of many things into one' (*quaedam collectio plurium ad unum*), one might assume that nonhuman animals gain cognitive access to universals by gathering together several singulars. For instance, a dog acquires the universal or the concept 'man' after having met several men.

However, this interpretation is incompatible with what Bacon says in his commentaries on Aristotle's *Metaphysics*. In both commentaries, he stresses that only humans engage in what he calls 'experimental cognition' (*cognitio experimentalis*), that is, a 'universal grasp of singulars' (*universalis acceptio singularium*) or the 'collation of singulars to one universal' (*collatio singularium ad unum universale*). Consequently, nonhuman animals only have a 'grasp of singulars not under some common nature' (*acceptio singularium non sub aliqua natura communi acceptorum*).[7] Now how can this be reconciled with what he says about the cognition of universals in the *Perspectiva*?

2 Köhler (2014), 360.

3 See Roling (2011), 245f., and Sobol (1993), 116f. Roling's interpretation is not entirely consistent as he claims that there is, for Bacon, 'a latent involvement of universals' ("latente Einbindung von Universalien") in nonhuman animal cognition but, nevertheless, a restriction 'to the domain of particulars' ("auf die Domäne der Einzeldinge"). Wood (2007), 42, in contrast, claims that "[n]o medieval suggests that animals can generalize." However, she gives no interpretation of Bacon's statement in the *Perspectiva*.

4 Hackett (2013), 235.

5 See *Perspectiva*, pars I, dist. 1, c. 4, ed. Lindberg (1996), 12.

6 See, for instance, *Perspectiva*, pars II, dist. 3, c. 9, ed. Lindberg (1996), 248. On Bacon's theory of internal senses see Dewender (2003), 155f., and Rignani (2006). On the claim that only very few medieval authors ascribed cogitation to nonhuman animals see Köhler (2014), 320–325.

7 See Roger Bacon, *Questiones supra libros prime philosophie Aristotelis I*, ed. Steele (1930), 8f., and *Questiones altere supra libros prime philosophie Aristotelis I*, ed. Steele (1932), 17–19. On those passages see also Hackett (2008), 127.

The easiest way to resolve the tension between these passages would be to say that Bacon changed his mind over time. While in his *Metaphysics* commentary, written at some point between 1237 and 1250, he still believed that non-human animals cannot engage in the cognition of universals, he substantially revised this view when writing the *Perspectiva*, at some point in or after 1260.[8] This is, however, merely speculative and calls for further explanation for why he changed his mind. Another option would be to argue that his statements in the *Perspectiva* are based on his realism about universals and his theory of so-called '*species*'.[9] As he explains in *De multiplicatione specierum*, the latter are produced by singulars and universals alike.[10] Therefore, one might assume that the way in which they are cognised is also similar. Yet even if this is correct it is not said that universal species can be cognised without intellectual faculties. Moreover, the tension with regard to what he says in his *Metaphysics* commentary still remains.

Accordingly, it makes sense to consider a third option. This option consists in saying that the sense in which Bacon applies the term 'universal' in the *Perspectiva* is a very special sense. More precisely, the universal Bacon is talking about in the *Perspectiva* is a so-called 'vague particular' (*particulare vagum*). That is to say, the dog does not form the concept 'man' after having met several men. Neither does it grasp the universal 'man' in terms of grasping the quiddity of man. Rather, it manages to discern between different kinds of particulars, namely, particulars which it knows very well, such as its owner or the neighbour's cat, from individuals of the same species which it has never seen before, such as a strange man or a strange cat. Now the perception of such strange or unknown individuals is vague insofar as the dog knows almost nothing about them. While his owner is *this particular* man for him or while the neighbour's cat is *this particular* cat, a strange man is just *some* man and a strange cat is just *some* cat. In other words, the dog distinguishes known from unknown particulars.

But then why are the latter called universals? In order to answer this question adequately it is necessary first to understand the context in which Bacon discusses the cognitive capacities of animals. This context is the distinction between three modes of seeing – a distinction he adopts from Alhacen's *De aspectibus*. This was one of the main sources for Bacon's own treatise on optics

---

8 On the chronology of Bacon's writings see Hackett (1997), esp. 316f.

9 On Bacon's realism see Maloney (1985).

10 See Roger Bacon, *De multiplicatione specierum*, pars 1, c. 2, ed. Lindberg (1983), 40–42.

and so he adopts several of Alhacen's terms and distinctions.[11] According to Alhacen, the first mode of seeing consists in perceiving things 'by the sense [of sight] alone' (*solo sensu*), that is, without the help of any other faculties of the soul.[12] In this case, one perceives light and colours universally but no distinction is made between distinct colours, let alone between distinct objects. The reason is that such distinctions require what Alhacen calls 'the power of discerning' (*virtus distinctiva*). It is by means of this power that we are capable of distinguishing between all kinds of things, and this is what constitutes the second mode of seeing.[13]

In Bacon's words, 'this mode of cognition is not only about light and colour, but about all things in which we distinguish a universal from a particular and particulars from each other'.[14] However, he specifies what he means by 'universal' in this context by explaining that it is a 'vague particular' (*particulare vagum*). A vague particular is, for instance, *some* colour or *some* man, which needs to be distinguished from particulars that are not vague, such as *this* colour or *this* man.[15] In Bacon's view, the distinction between vague and non-vague particulars is based on memory and works 'by a comparison of things we see to those things we have already seen before' (*per comparationem rei vise ad eandem prius visam*). For once we remember that we have seen the man we currently see before us, we no longer simply see *some* man, that is, a vague particular, but *this* man, namely, the concrete person we have seen before.[16]

The crucial point of Bacon's explanation now is that this kind of cognition is found at the level of the sensory soul. For many Latin perspectivists, this was

---

11 On the reception of Alhacen in the Latin West see Lindberg (1976), 104–121; Smith (2001), lxxx–xciv.

12 See Alhacen, *De aspectibus*, lib. II, c. 3, ed. Smith (2001), 98.

13 Ibid., 100f.

14 Roger Bacon, *Perspectiva*, pars I, dist. 10, c. 3, ed. Lindberg (1996), 154: "Et hic modus cognitionis non solum est circa lucem et colorem, sed circa omnes res in quibus distinguimus universale a particulari et particularia ab invicem."

15 See ibid.: "Sed non excluditur particulare vagum, nam illud est ita commune sicut suum universale, et convertitur cum eo, ut aliquis color, aliqua lux, aliquis homo, aliquis bos. Cognoscere igitur universalia ab invicem et a particularibus, et particularia ab invicem per comparationem rei vise ad eandem prius visam, recolendo quod prius fuerit visa et nota videnti, facit hic secundum modum comprehendi per visum."

16 Bacon's description very much reminds one of Averroes' account of memory because, according to Averroes, "to remember something is not primarily to recognize it as past object of perception but to comprehend it as this particular thing," as Black (1996), 173, puts it. However, Bacon very likely adopted the theory of vague particulars not from Averroes but from Avicenna. On Avicenna's theory and its reception in the Latin West see Black (2011). On vague particulars in Bacon see also Maloney (1985), esp. 812.

doubtful because the second mode of seeing was usually referred to as 'cognition by knowledge' (*cognitio per scientiam*).[17] And since 'knowledge' was considered to require intellectual powers, the second mode of seeing was taken to depend on intellect and reason.[18] Bacon, however, rejects this view. One of the strongest points in support of his argument is the cognition of nonhuman animals. For even though such animals lack an intellectual soul, they nevertheless engage in the second mode of seeing: dogs and many other animals obviously discern things they have seen before or 'of which they have memory' (*quarum habent memoriam*) from those things they have not seen before. This they do by virtue of the cogitative power which Bacon takes to be identical with Alhacen's *virtus distinctiva*.[19]

Since Bacon explicitly states that this distinction extends to both particulars and universals it is, of course, correct to say that he takes nonhuman animals to possess 'some form of universal cognition', as Köhler put it.[20] Still, one needs to specify what is meant by 'universal cognition' in this context. Obviously, the dog does *not* cognise the universal 'man' if universal here means the quiddity or the universal essence of man. As Bacon stresses in his commentary on the *Metaphysics*, the dog is incapable of acquiring the universal 'man' through the repeated perception of several individual men. But it does cognise the universal 'man' if universal is understood as a "generic universal," as Jeremiah Hackett calls it.[21] This might seem like a terminological form of cosmetic repair. Still, it makes sense to speak of universals here for a couple of reasons. First, seeing *some* man is much closer to cognising the universal 'man' than is seeing *this* man. The cognition of *this* man is much more concrete than the cognition of *some* man. This is why Bacon states that a vague particular is 'as general as its universal and can be converted with it' (*est ita commune sicut suum universale, et convertitur cum eo*).[22] Second, Aristotle also applies the term 'universal' in

17 See Roger Bacon, *Perspectiva*, pars II, dist. 3, c. 9, ed. Lindberg (1996), 246.

18 See Bacon, *Perspectiva*, pars II, dist. 3, c. 9, ed. Lindberg (1996), 246.

19 On Alhacen's influence on Bacon regarding the cogitative power see Summers (1987), 161; Tachau (1988), 9–11; Dewender (2003), 155; Rignani (2006), 1227f.

20 See n2 above.

21 See Hackett (2008), 130. One could also argue that what Bacon has in mind here is nothing but the kind of 'first universal' (*proton katholou*) of which Aristotle speaks in *Analytica Posteriora* II.19, 100a16. Consequently, one could also speak of 'primitive universals', 'rudimentary concepts', or 'pre-predicative sub-universals', as do some interpreters of Aristotle; see Steiner (2005), 68f.

22 See n15 above. Bacon is not the only one who claims that vague individuals are convertible with universals. Maierù (2011), 361–363, mentions a number of authors who worked in Paris at about the same time and who held the same views concerning vague individuals.

this sense at the beginning of the *Physics*.[23] This passage has puzzled (and still puzzles) many interpreters but it does provide some grounding for a theory of universals as vague particulars.

Nevertheless, there still remains a chasm between the sensory and the intellectual realm, thus between human and nonhuman animals because, in the end, the dog will never cognise anything other than particulars, no matter what degree of precision its cognition has. Its cognition of particulars amounts to more than a simple act of recognition because it is capable of discerning individuals of one kind from those of another kind or one individual of a kind from another individual of the same kind. Yet, like Aquinas and Albertus Magnus, Roger Bacon thinks that it does not apprehend individuals 'under some common nature' (*sub aliqua natura communi*).[24] Therefore, his position is obviously less radical than it might seem at first glance.

---

23 See Aristotle, *Physica* I.1, 184a24-b14.

24 See n7 above. Therefore, I find it rather misleading to say that, according to Bacon, "animals can distinguish universal classes from one another" or "recognize and distinguish universal categories and individuals," as Sobol (1993), 116, puts it.

CHAPTER 12

Universal Desire and Experience (John Buridan)

In the previous chapters, the discussion of universal cognition and concept formation in nonhuman animals has largely centred on the question of how these animals perceive what they presently perceive. That is to say, the question was, for instance, how a sheep perceives a wolf or how a dog perceives a stick. More precisely, it has been asked whether their perception of the world involves conceptualisation, categorisation, and generalisation. A further question that has been mentioned in the introductory section to this part concerns the way in which nonhuman animals identify objects. This question has already been touched on by some of the authors discussed above. For instance, Pseudo-Peter of Spain considers the possibility that sheep identify individual wolves by using a common form of wolf. Yet, what is common to all these examples is that they focus mainly on how a *present* stimulus is (mentally) represented by the cogniser: as a coloured pattern, as an object, as a member of a kind, and so forth? But what about those situations in which no stimulus is present?

Take the example of a donkey that is looking for food. In this situation, the animal has a particular desire. But how does it find what it desires? In other words, how does it (mentally) represent the object or thing it is looking for? Would we not be inclined to say that the donkey's search for food is based on a *universal* or *general* mental representation of food rather than on a particular representation? Admittedly, the donkey might be looking for the particular piece of bread its owner had put in its jacket in front of its eyes a few minutes ago. In this case, its search is likely based on the mental representation of *this particular* piece of bread just like our search for the car key, for example, would be based on a mental representation of *this particular* key. But suppose the donkey simply looks for any piece of bread. Then its search is likely based on a general representation of bread just like our search for any key would be based on a general representation of key. One could, of course, raise an immediate objection to this argument, namely, that any kind of search is based on the representation of a *particular* thing. If one is looking for *any* key one might still have a *particular* key in mind and then look for objects which are very similar to this particular key. In this case, the success of the search does not require a general mental representation. Instead, it depends on the capacity to cognitively establish similarities between a particular thing that is mentally represented and a thing that is 'out there' in the world, so to speak.

 | DOI 10.1163/9789004363779_016

How did late medieval authors dealt with such cases? Did they ascribe general mental representations to nonhuman animals or did they try to find alternative explanations? One might immediately assume that a good place to look for an answer is medieval commentaries on Aristotle's *De motu animalium*, because in this text Aristotle examines the physiological as well as the psychological processes underlying the movements of animals. In Chapter 7, in particular, he describes how we are usually moved towards something drinkable:

> "My appetite says, I must drink; this is drink, says sensation or imagination or thought, and one immediately drinks. It is in this manner that animals are impelled to move and act, the final cause of their movement being desire; and this comes into being through either sensation or imagination and thought."[1]

The process described here apparently involves three main stages. There is, first, the stage of *appetite*, that is, the stage in which we feel thirst. This is, second, followed by the *identification* of what could assuage our appetite, namely, a drink. Third, there is the *action* of assuaging the appetite: the action of drinking.[2] For the present purpose, the first and second stages are the most interesting stages because in those stages mental representations come into play (supposing that they come into play at all). However, Aristotle himself neither says anything about the nature of the mental representation involved here nor makes explicit whether his description applies to all animals.[3] All he says is that the identification of what is searched for happens by 'sensation or imagination or thought'. This, however, leaves open *how* the mental representation on which the search is based looks like and *who* has it. Even if one were to argue that it is the faculty of imagination which guides the search in nonhuman animals, the above-mentioned question concerning the generality of such representations still remains, and it is rarely discussed in medieval commentaries on *De motu animalium*.[4] Nevertheless, there are other texts containing discussions of the example of an animal that is looking for drink or food which attend to the aspects of particularity and universality.

1 Aristotle, *De motu animalium* 7, 701a32-36, tr. Forster (2006), 463.

2 For a detailed discussion see Labarrière (2004), 204–214.

3 The latter aspect is also stressed by De Leemans (2005), 144f. See also De Leemans & Klemm (2007), 169.

4 John Buridan, for instance, simply takes the example to illustrate that we sometimes do things without waiting for an intellectual judgment about whether or not this is a good thing to do; see John Buridan, *De motibus animalium*, eds. Scott & Shapiro (1967), 545. This is also what Labarrière (2004), 212, takes to be the original function of this example.

One philosopher who pays some attention to this question is John Buridan. In one version of his commentary on Aristotle's *Physics*, for example, he states that 'a horse or a dog desires *in a universal mode* when being hungry or thirsty'.[5] This is no trivial claim, since the appetite Buridan is talking about is a 'sensitive appetite' (*appetitus sensitivus*). This appetite is 'extended and material like the senses' (*extensus et materialis sicut sensus*). Therefore, Buridan argues, universality seems to be nothing exclusively intellectual, but can be found at the level of sensory powers, too. But is it actually true that sensitive desire or appetite is universal? For Buridan, nonhuman animals are a good example in this context. For instance, if a horse is looking for food or water it does not desire *this* or *that* particular portion of water or oats but just *any* portion. In other words, it desires 'indifferently' (*indifferenter*). Of course, there might be cases in which a horse or another animals looks for some particular thing. But, usually, its desire is indifferent or universal. Admittedly, to claim that indifference is the same as universality might seem odd. For present purposes, however, the more important question is what the universality Buridan is talking about amounts to. More precisely, the question is: is the animal's appetite based on a mental representation, and if so, can one infer from the universal nature of the *appetite* to the universal nature of that *representation*?

At first glance, it seems that, in Buridan's view, one cannot make such an inference. One sign of this is that a desire is also universal if the agent (or thing) having this desire has no cognition at all. Buridan gives the example of fire which has the 'desire' to burn things. Since this happens indifferently, too – fire burns any inflammable thing just as a horse usually eats any edible thing – the fire's desire can be said to be 'in a universal mode' (*modo universali*) as well.[6] Consequently, no cognition is needed in order to have a universal desire. Hence, no mental representation is involved either. Otherwise, a non-cognitive agent like fire could not be said to desire in a universal mode. This point is emphasised by Buridan in his commentary on Aristotle's *Metaphysics*. Here he says that such appetites 'are not mediated by cognition' (*non sunt mediante*

5 John Buridan, *Quaestiones super octo libros Physicorum Aristotelis* (*secundum ultimam lecturam*), lib. I, q. 7, eds. Streijger & Bakker (2015), 66: "Tertio, quia appetitus sensitivus ita est extensus et materialis sicut sensus; et tamen equus et canis per famem et sitim appetunt modo universali. Non enim hanc aquam vel avenam magis quam illam sed quamlibet indifferenter; ideo quodcumque eis praesentetur, bibunt eam vel comedunt." For other texts in which Buridan mentions this example, see Pluta (2002a), 31–40.

6 See ibid.: "Et etiam intentio, potentia vel appetitus ignis ad calefaciendum est modo universali, scilicet non determinate ad hoc lignum, sed a quodlibet lignum calefactibile indifferenter, licet actus calefaciendi determinatur ad certum singulare; et ita etiam potentia visiva est modo universali ad videndum."

cognitione).[7] However, one ought to be careful regarding the connection between the universal nature of appetites and the nature of mental representations because, as has been said, Buridan's point is that no cognition is needed in order to have a desire. That is, cognition does not *necessarily* precede a desire. But it is *possible*, of course, that cognition comes along with a desire, namely, in those agents which do have cognitive powers, such as horses or humans. And so if their desire is universal must not their mental representations be, too?

According to Olaf Pluta, this is indeed the conclusion one must draw. In his opinion, "Buridan clearly affirms that animals, and dogs in particular, can refer to things universally [and] [t]his also means that they can mentally represent a particular bowl of water as a member of the class or universal category 'water'."[8] Of course, representing a particular as a member of a class or universal category is not entirely the same as having a universal representation of a thing, as we have seen above. Yet, for Pluta, "Buridan maintains that animals are capable of universal reference in the realm of conceptual knowledge."[9] 'Conceptual knowledge' means both "the ability to represent mentally an individual object of sensation as a member of a class or universal category, or, conversely, to signify a plurality of individual objects by a single mental entity." The latter meaning is what is at issue in the present chapter. For even if one holds that the horse does not need the *concept* of water for finding water, it still seems to need "a single mental entity" by means of which it can cognitively refer to or identify "a plurality of individual objects," as Pluta puts it.

From a systematic perspective, Pluta's point is fairly reasonable. However, it is also fairly reasonable to doubt that his interpretation is the only possible (let alone correct) interpretation of Buridan. Two main points militate against Pluta's interpretation. The first is more general and systematic in nature and concerns the actual inference from the universality of desire to the universality of mental representations. According to Pluta, it is correct to make such an inference. The question is, however, whether it is *necessary* to make this

7 John Buridan, *In Metaphysicen Aristotelis Quaestiones*, lib. I, q. 7, Paris (1518), fol. 7rb: "Sed est dubitatio de appetito sensitivo quomodo actus appetendi sit circa universale cum actus sentiendi <non> sit circa universale. Ad hoc videtur mihi dicendum quod talis appetitus magis sunt naturales quam animales, ut sitis et fames, et non sunt mediante cognitione, immo immediate a natura propter carentiam nutrimenti. Sed ad persequendum obiectum appetitus exigitur sequens cognitio vel memoria, non tamen ideo quod ille appetitus fiat mediante cognitione illa vel memoria. Modo tales appetitus naturales non sunt actus distincti ab illis potentiis appetitivis naturalibus: ideo sunt circa universalia sicut dictum est quod omnes potentiae sunt circa universalia." On the addition of '*non*' see Klein (2016), 102, n. 248.

8 Pluta (2015), 280.

9 Ibid., 278.

inference in order to coherently account for the behaviour of nonhuman animals in Buridan's theory. One could, for instance, argue that since the appetite of a horse (i.e. of a cognitive agent) is very much the same like the appetite of fire (i.e. of a non-cognitive agent), there is no cognition involved in the case of the horse just as there is no cognition involved in the case of fire. In short, the horse has a universal *appetite* for water since it is going to take *any* portion of water. But it does not have a universal *mental representation* of water because its appetite can be assuaged without cognition, just as fire needs no cognitive powers in order to burn something. In modern animal psychology, the methodological principle behind this kind of argument is generally known as 'Morgan's Canon'. According to this principle, one should avoid explaining animal behaviour by referring to higher psychological processes if it can also be explained by lower processes.[10] If one takes the appetite of fire to be such a lower process, one can apply Morgan's Canon to the appetite of the horse and thus explain it without reference to any kind of cognition.[11]

One might, of course, reject this explanation by saying that the horse has cognitive powers and will thus use them to find nutrition, hence possesses certain mental representations of what it seeks. Still, there is some reason to doubt that those mental representations are as universal as Pluta takes them to be. The reason is that Buridan does not actually aim to provide evidence for universal mental representations at the level of the sensory soul. Rather, his aim is to show that "the materiality of sense is not the ground of the singularity of sensitive cognition," as Peter King has put it.[12] That is to say, Buridan has no doubt that sensitive cognition is singular. What he doubts is rather that the singularity of sense perception can be proven with reference to the material nature of the senses.[13] In his view, *all* powers – no matter whether they are material or immaterial – 'are about universals' (*sunt circa universalia*). The reason is simply that all of them are indifferent with regard to their objects: sight indifferently perceives all colours and the desire to eat something is indifferent towards edible things. Yet, this *being about universals* does not amount

---

10 The original formulation of this principle is found in Morgan (1894), 53: "In no case may we interpret an action as the outcome of the exercise of a higher psychical faculty, if it can be interpreted as the outcome of the exercise of one which stands lower in the psychological scale." For a recent critical discussion see Starzak (2014), 12–32.

11 Biard (2012), 59, seems to go in this direction by saying about Buridan's theory that "l'appétit n'est pas le sens en tant que puissance cognitive, qui quant à lui appréhende toujours telle chose matérielle singulière."

12 King (2001), 12.

13 The most recent and thorough reconstruction, analysis, and discussion of Buridan's argument is provided by Klein (2016), esp. 95–105.

to *cognising universals* but rather resembles what has been called *cognising universally* in previous chapters of this part (see Chapters 8 and 9). One could also say that all powers are universal under the aspect of intentionality.[14] Still, this does not pave the way for the passage from universal appetite to universal mental representations. But is this to say that Buridan denies universal cognition to nonhuman animals entirely?

If one looks at his commentary on the *Prior Analytics* it seems as if the contrary is the case because in one passage of this commentary he brings up the famous example of the dog that begins to fear sticks or stones and he argues that the dog's reaction of flight is based on what he calls an 'experimental judgment' (*iudicium experimentale*). Why Buridan ascribes a *judgment* to the dog here is the subject of Chapter 17. For the moment, it suffices to concentrate on the aspect of experience, because this is where universal cognition comes into play. According to Buridan, the dog behaves like a human being that experiences fire as hot and learns to stay away from it. For instance, an inexperienced child who sees a fire A and touches it it will sense that this fire is hot. It might have the same experience with another fire B. But after a couple of painful encounters with fires the child might finally arrive at the conclusion that any fire, or fire in general, is hot. Thus, if it then sees some fire C it will – by experience – judge this to be hot and stay away. Similarly, the dog stays away from sticks because it knows by experience that sticks are (or can be) painful.[15]

As discussed in the previous chapters, this can be taken as evidence for the fact that the dog is endowed with the capacity of generalisation, because it seems to know that sticks *in general* are painful. Buridan, however, seems to favour a more conservative conclusion here. The first thing he emphasises is that the principles providing the basis for such experimental judgments are

---

14 See King (2001), 12, n. 30, who suggests that "the intentional nature of desire introduces a kind of opacity (intentionality produces intensionality): the horse wants some-water-or-other, which cannot be identified with any particular water, but is such that any particular water satisfies it."

15 See John Buridan, *Quaestiones in Analytica Priora*, lib. II, q. 20a, ed. Hubien [undated typescript]: "Verbi gratia, si tu ad sensum cognovisti quod ignis A erat calidus, et postea idem de igne B, et sic de multis aliis, tu postea videns ignem C, et non tangens ipsum, judicabis per memoriam de aliis et propter similitudinem quod ille ignis C est calidus; et hoc non est, proprie loquendo, judicium per sensum, quia non tangis ipsum, nec solum per memoriam, quia memoria proprie non est nisi prius cognitorum et tamen ipsum ignem C numquam alias vidisti nec cognovisti; sed hoc judicium vocatur 'experimentale'. Et non solum homines, immo aequaliter brutae hujus modi judicio utuntur; unde propter hoc canis timet lapidem si aliquis laesit ipsum. Et omnia praedicta principia sunt singularia, et sunt principia in arte vel in prudentia, et non in scientia speculativa vel demonstrativa."

'singular principles' (*principia singularia*). That is to say, the dog's reaction is based on the principle that *this particular* stick is painful because *that particular* stick it has sensed before had been painful, too. In other words, it is incapable of fully abstracting from various particular encounters with sticks and so it does not arrive at the universal principle 'All sticks are painful'.[16] Second, Buridan also remarks that the faculty that judges in this situation is not the intellect, but the estimative power, which is why nonhuman animals have such cognition.[17] Admittedly, this argument might be taken to be insufficient because it rests on the assumption that the estimative power cannot access the realm of universals. Although this was taken for granted by Buridan and most other medieval thinkers, one could ask for some evidence for the fact that estimative perception is restricted to particulars. The easiest way to address this objection is to refer to 'Morgan's Canon' once again. So as long as one can coherently account for the dog's behaviour without reference to universal principles or generalisation, one should stick to this explanation. But is Buridan able to give such an explanation?

The short answer is 'yes', because according to Buridan if one judges by experience that a given fire C is hot, one judges 'by memory of other [fires] and because of [some] similarity" (*per memoriam de aliis et propter similitudinem*).[18] Put differently, in order to see that *this* particular fire C is hot, I do not need to see that *all* fires are hot. Rather, it suffices to see that this fire C is *similar* to that fire A or B. Likewise, the dog need not realise that *all* sticks are hurtful. Instead, it only needs to recognise that the thing it sees is similar to what it has seen before, just as Albertus Magnus' weasel recognised snake bites and endive leaves (see Chapter 8). The recognition of similarities might, of course, require some degree of abstraction because a given fire or stick A is not entirely identical to another fire or stick B. However, this kind of abstraction need not go as far as the abstraction of the concept of fire or stick. Whether it must lead to the abstraction of some sort of common form or some general mental representation of stick is worthy of consideration. But as has been shown above, Buridan seems to be rather reluctant to grant such representations to nonhuman animals.

---

16 This is also stressed by Klein (2016), 178. Similarly, Pluta (2015), 276, notes that this kind of knowledge "is, however, particular in the sense that it refers to a particular sensation and memory or associates a series of particular sensations and memories."

17 See John Buridan, *Summulae de demonstrationibus* 8.5.4, ed. de Rijk (2001), 128f.: "Et non fit hoc per intellectum, quia sic faceret iuvenis catulus vel caniculus, licet non habens intellectum; sed hoc iudicat virtus aestimativa."

18 See n15 above.

CHAPTER 13

General Mental Representations (Peter of John Olivi)

In contrast to Buridan, Peter of John Olivi seems explicitly to allow for mental representations that go beyond the level of particularity in nonhuman animals. In his commentary on the *Sentences*, he also discusses the example of thirst. In this context, he points out that there is a difference between the so-called 'species' we acquire when we perceive something and those species by means of which we imagine, remember or think of something. For instance, if there is a glass of red wine standing in front of me, the species I perceive is the species of *this particular* glass of red wine. In other words, I mentally represent this particular glass of wine and so the mental representation I have in this moment is of a very particular individual. The situation is different, however, if I feel the desire or intention to drink some wine when there is no particular glass in sight. In this case, I mentally represent wine 'in general or universally' (*in generali seu in universali*), according to Olivi.[1]

As Juhana Toivanen emphasises, it is important to distinguish between universality and generality here. The wine I imagine is not *universal* wine because the mental representation of wine which I have in my imagination "does not represent the universal essence" of wine. In other words, I do not mentally represent 'wineness' or the quiddity of wine. Rather, the mental representation I have is the representation of what Toivanen calls "generic" or "general" wine.[2] It is generic or general in the sense that it enables me to find *any* glass of wine. If, instead, I were to mentally represent a particular glass of wine, my search would only be successful if I found *this particular* glass. Thus, it seems as if Olivi holds the view that general mental representations are required if one tries to find something that is not currently being perceived.

1 See Peter of John Olivi, *Quaestiones in secundum librum Sententiarum*, q. 36, ed. Jansen (1922), 634: "[...] tum quia quando hoc intendimus vel praecogitamus, nunquam hoc praecogitamus per species per quas res videri possunt, sed solum per species memoriales per quas res absentes possumus imaginari vel rememorari vel recogitare. Et ita non oportet quod visio rei praecedat rei, quia aliquando hoc non intendimus nisi in generali seu in universali; utpote volens videre asinum vel emere vinum non oportet quod praecogitem in particulari hunc vel illum asinum vel hoc vel illud vinum, sed sufficit quod in generali."

2 Toivanen (2013a), 323f.

 | DOI 10.1163/9789004363779_017

Still, two questions can be raised with regard to Olivi's explanation. The first question is whether his account of general mental representations applies to nonhuman animals too since the example he gives obviously concerns human beings. According to Juhana Toivanen, this question is legitimate, but from the context it is clear that what Olivi says about the nature of human mental representations in such situations also applies to the representations of other animals because the process described here takes place at the level of the sensitive soul.[3] Of course, one could object that there might be a fundamental difference between sensory cognition in humans and nonhuman animals: while we acquire universal mental representations by virtue of our intellect, nonhuman animals do not, and so the general representation of wine in our imagination might be some kind of derivative representation of the universal representation of wine we have in our intellect whereas this is not the case for other animals.

This is exactly where the second question arises, namely the question of how one can conceive of general mental representations in those animals that do not have intellectual powers. In order to answer this question, it will be helpful to adopt the above-mentioned distinction between *universal* and *general* mental representations. The crucial point is in what sense they differ from *singular* representations. As has been said, having a *universal* representation consists in representing the quiddity of a thing which is common to all things of the same kind. A universal representation is usually taken to be immaterial because it abstracts from the material circumstances that make a thing *this particular* thing. Having a *singular* representation, by contrast, means to represent such a particular thing only. Singular representations might, of course, be necessary for acquiring universal representations, depending on one's theory of concept formation. Nevertheless, they are the exact opposite of universal representations, so to speak.

General representations are in some sense half-way between universal and singular representations because, on the one hand, they are material representations, like singular representations, but on the other hand, they are unlike singular representations in that they represent more than just one particular thing. As an illustration of what such a general representation might look like, one can, once more, turn to the example of red wine: while my universal representation of red wine is a totally abstract and immaterial representation and my singular representation a totally concrete and material representation, my general representation might be something like a patchwork made of several singular representations of red wine. It is thus material (like my singular and

3 See Toivanen (2013a), 323.

unlike my universal representation), but not singular (like my universal and unlike my singular representation).[4] Admittedly, it remains unclear to some extent whether this is what Olivi has in mind when speaking of general representations, especially because he does not explicitly make a terminological distinction between general and universal representations in this passage. But supposing that he accepts this distinction, he would be a good – and, among the thinkers covered in this part, a relatively unique – example of a medieval philosopher who grants general mental representations to nonhuman animals. To be clear, Olivi does not stand out from other authors insofar as he does not cast doubt on the lack of intellectual powers in nonhuman animals either. As regards the aspect of universal cognition and concept formation, however, he does provide an interesting explanation of how nonhuman animals can, nevertheless, reach beyond the level of mere particularity in some sense.

As has become clear in this part, many other thinkers had an interest in this question, too, as they obviously wondered whether the lack of intellect and reason really implies that nothing but particulars can be perceived. While some were rather sympathetic to making this connection, others considered in what sense this connection might be too strict. In the end, none of them ascribed the cognition of universals in terms of quiddities to nonhuman animals and none of them held that nonhuman animals engage in concept formation as we do by virtue of intellect and reason. Nonetheless, all of them paid attention to whether the distinction between particular and universal cognition is as clear-cut as it seems to be or whether there are shades of grey, so to speak. In the following part, we shall examine whether such nuances also exist with respect to the second intellectual operation, namely, judging. Because if nonhuman animals are incapable of forming concepts, the question is then: are they nevertheless capable of forming judgments?

4 A very similar description of such representations is given by Sorabji (1993b), 63. Although he does not talk about Olivi, but Avempace (who deals with Galen) in this context, the issue is the same. However, it is rather unlikely that Avempace's account somehow influenced medieval thinkers in the Latin West. For a very concise discussion of Avempace's position see also Druart (2016), 78f. A complete translation of the passage in question can be found in Druart (1980), 76. I have also profited from an unpublished paper on medieval Islamic theories of animal cognition which was kindly sent to me by Thérèse-Anne Druart.

PART 3

Judging

Introduction to Part 3

Nonhuman animals are capable of doing many things. They recognise people, other members of their species, or nonliving beings. They discriminate between individuals and members of different kinds, and they are able to see similarities between various objects. None of the thinkers covered in the previous part cast doubt on their ability to do this. However, many of them stress that nonhuman animals do not recognise and distinguish between things by using general terms or concepts. That is, sheep or dogs do not actually understand that there is something like kinds or categories, and when a sheep sees a wolf it does not, strictly speaking, perceive the wolf as falling under the concept of wolf. Moreover, many thinkers emphasise that sheep and other animals do not detect similarities to the extent that they cognise universals in terms of quiddities. This, they maintain, remains the privilege of intellect and reason and thus sets humans apart from other animals. To be sure, some of them hold that other animals have an idea of kinds as well. Some of them also grant them some sort of general mental representations. Still, no thinker went so far as claiming that the cognition of universals and concept formation can be achieved by the sensory powers, which , according to Aristotelian psychology, we share with other species.

It therefore seems impossible that nonhuman animals can engage in judging because, according to what is sometimes called the "building-block model" (see Chapter 6) judgments are propositions. Propositions, in turn, are made of general terms or concepts.[1] So any being which lacks concepts, also lacks the capacity to form propositions or to have "propositional attitudes," as Donald Davidson famously put it.[2] Propositional attitudes include all kinds of attitudes, from beliefs and opinions to hopes and fears. Quite often, such attitudes are summarily called 'thoughts'. In every case, they have a propositional structure, in the view of Davidson and other modern philosophers because to have, for instance, a belief means to 'believe *that* X'. Now one can, of course, criticise this account of judgments for various reasons. But this does not detract from the fact that the medieval thinkers covered by this study took judgments to be propositions and propositions to be made of concepts.[3] Thus, the sheep

---

1 On this model see, for instance, Glock (2010), esp. 15–19.

2 See Davidson (1982). A concise summary of Davidson's account provides Wild (2013), 90–104.

3 For a brief overview of medieval theories of judging see Tachau (1993) and Perälä (2014). The most comprehensive and insightful studies are still Nuchelmans (1973), (1980); Perler (1990), (1992).

 | DOI 10.1163/9789004363779_018

cannot strictly speaking *judge* that this wolf is hostile or that this lamb is its offspring unless it forms such concepts as 'wolf', 'hostile', 'lamb', 'offspring', and so forth.[4]

This makes it all the more surprising that various thinkers nevertheless ascribed the capacity to 'judge' (*iudicare*) to nonhuman animals in various respects: sheep judge wolves to be dangerous, dogs judge sticks to be painful, or swallows judge straws to be useful, to name just a few examples. The crucial question, of course, is whether the term 'judging' is applied to nonhuman animals in the same sense in which it is applied to cognitive processes in humans. Did medieval authors ascribe judgments in terms of propositions to other animals? Or did they have a broader concept of judging which also covers non-propositional forms of judgments? Or were other animals said to judge only in a metaphorical way of speaking? And if both human and nonhuman animals are capable of judging, what is the difference between the way in which humans judge and the way in which sheep, dogs, swallows, or other animals judge? Do such nonhuman animals judge deliberately or instinctually? Are they capable of reflecting on their own judgments and of discriminating between correct and incorrect judgments, for instance?

These are just some of the questions that will be addressed in this part. As in the previous part, examples will be given for answers to each of these questions. But it is important to note that not each and every thinker covered in the following discussion provides answers to each and every question. Rather, they mostly focus on single aspects, which are determined by their specific motivation for discussing the capacity of judging in nonhuman animals. The structure of this part therefore follows once more a systematic, rather than a chronological arrangement. It will begin with a brief chapter on the idea of sensory judgments as found in Aristotle and further developed in Avicenna's theory of estimation (Chapter 14). Chapters 15–17 deal with how this idea was received by thirteenth- and fourteenth-century thinkers in the Latin West, namely, Thomas Aquinas, Albertus Magnus, John Buridan, and Peter of John Olivi. All of them claim that nonhuman animals are capable of judging in a way, but all of them also hold different views on what this capacity amounts to. But not every author accepted such a concept of judging. Most prominently, William Ockham's secretary Adam Wodeham argue for a narrower concept. His interesting debate with Gregory of Rimini will be the topic of Chapter 18.

4 A relatively early but illuminating discussion of this argument is found in John Blund's early 13th-century *Tractatus de anima*, c. 19, §§258 and 260, eds. Callus & Hunt (2013), 140.

CHAPTER 14

The Idea of Sensory Judgments

In order to understand how medieval authors came to say that, for instance, the sheep *judges* the wolf to be hostile although it lacks the concepts of wolf, hostility, and so forth, it is necessary to look briefly at the ancient and early medieval roots of the idea that there exist not only intellectual judgments but also "judgments of sense."[1] What does it mean to say that the senses judge? According to Aristotle, the external senses are "distinction-making faculties," as Martha Nussbaum puts it.[2] That is, they distinguish between different kinds of sensibles. For instance, our sense of sight distinguishes yellow from green and our sense of taste bitter from sweet. The task of discrimination is not solely a sensory task, of course, because both the intellect and the power of imagination make certain distinctions, too. But each of these powers distinguishes different objects. While the external senses discern sensible things, the intellect discerns concepts and the imagination discriminates phantasms. Admittedly, there is a substantial difference between distinguishing concepts and distinguishing colours or flavours. The former is a much more complicated cognitive operation, one might say, and no sense could ever perform this task because it has no access the realm of concepts. Still, the basic operation of distinguishing is the same and this is why Aristotle calls the intellect, the imagination, and the sensory powers 'distinguishing' or 'discerning faculties' (*kritika*).[3]

One could object that distinction-making is not the same as judging because distinction-making can be found in non-living beings as well. Take, for instance, a magnet. It clearly distinguishes various things, such as iron and gold. However, nobody would grant the magnet cognitive powers, let alone the capacity of judging. Similarly, sensory discriminations need not qualify as judgments. Therefore, on this argument, one should refrain from translating the Greek term '*krinein*' as 'judging'.[4] This is a legitimate objection indeed, but with respect to medieval theories it misses the point for a very simple reason: in the Latin translations of Aristotle, the term '*kritika*' was usually rendered

---

1 I borrow this expression from the very valuable study by Summers (1987). Similarly, Garceau (1968), 241–251, speaks of the "jugement des sens." On the medieval reception of this idea see also Tachau (1993).

2 Nussbaum (1985), 232.

3 See Aristotle, *De motu animalium*, c. 6, 700b19-22.

4 See, for instance, Tellkamp (1999), 141–144,; Davids (2017), 151. See also Summers (1987), 21–28.

 | DOI 10.1163/9789004363779_019

with '*iudicativa*' and the corresponding verb '*krinein*' with '(*di*)*iudicare*'.[5] Thus, even if one argues that 'judging' is not a good English translation of the original Greek term, it does not do away with the fact that Latin authors were used to speaking of 'judgments' in this context, even long before Aristotle's writings on psychology were translated into Latin. For instance, Augustine already points out that the external senses as well as the so-called 'inner sense' (*sensus interior*) 'judge somehow about corporeal objects' (*de corporibus quodam modo iudicare*), both in human and nonhuman animals.[6]

The idea of judgments by the inner senses was fleshed out in particular by Avicenna, in connection with his theory of estimation, which became highly influential in the Latin West, as we saw in the previous part. According to Avicenna, both 'the common sense and the external senses discern in a way and judge because they say that this moving thing is black and this red thing is sour'.[7] That is to say, our senses do not simply receive sensory inputs, but they also discern and combine what they perceive. The faculty of estimation is especially relevant in this context and it is even more relevant for explaining nonhuman animal behaviour. Avicenna calls it the 'ruler' (*domina*) of the animal soul, and he stresses that what it does is 'delivering a judgment' (*iudicans iudicium*).[8] Why does it make sense to speak of judgments here, especially since Avicenna emphasises that an estimative judgment differs from an intellectual one in that the former is 'not definite' (*non definitum*)? Moreover, since he also stresses that the estimation is 'a more excellent judge in animals' (*excellentior iudex in animalibus*), on the one hand, but prone to error, on the other hand? For instance, if one sees a yellow liquid, he says one can easily believe it

5 In the Aristoteles Latinus, *De motu animalium* (*transl. Morbeka*), c. 6, 700b19f., ed. De Leemans (2011), 55, the above-quoted passage thus reads: "Et enim fantasia et sensus intellectui eundem locum habent: iudicativa enim omnia."

6 Augustine, *De libero arbitrio*, lib. II, 5.12, §§49–50, ed. Green (1970), 245. On Augustine's account of an inner sense and the role it plays in nonhuman animal cognition see Brittain (2002), 288–295; Heller-Roazen (2007), 131–141. On the history of the idea of (common) sense judgments see also Heller-Roazen (2008).

7 Avicenna Latinus, *Liber de anima*, lib. IV, c. 1, ed. van Riet (1968), 6: "Sensus vero communis et sensus exteriores discernunt aliquo modo et diiudicant: dicunt enim hoc mobile esse nigrum et hoc rubicundum esse acidum."

8 Ibid., 8: "Et haec virtus [i.e., aestimatio; A.O.] sine dubio consistit in nobis; quae est domina, iudicans in animali iudicium non definitum sicut iudicium intellectuale, immo iudicium imaginabile coniunctum cum singularitate et forma sensibili, et ex hac emanant quamplures actiones animalium." On estimation and estimative judgments in Avicenna see Black (1993), 227ff.; Black (2000), 61; Di Martino (2008), 36f.; Tellkamp (2013), 211.

to taste sweet and thus to be honey although it is not honey, but bitter tasting bile.[9] But if this can only be corrected by an intellectual judgment, why should one call estimative judgments 'judgments' and not simply 'perceptions'?

The main reason that it makes sense to speak of judgments here is that, according to Avicenna, the estimative power elicits intentions. In other words, it turns the raw material delivered by the senses into meaningful information.[10] For instance, when a sheep sees a wolf, it primarily perceives sensory forms: the wolf's colour, its odour, its shape, and so forth. But this means nothing to the sheep unless the estimative power somehow *evaluates* these pieces of sensory information and then tells the sheep whether what it sees is, for example, something dangerous or harmless. That is, the estimative power informs the sheep about whether it needs to run away or not. Surely, the sheep's reaction of flight is instinctual, as has been said in Part 2. It is incapable of not running away in a situation in which it has perceived a wolf. Still, the theory of estimation can plausibly explain why there are different reactions to similar stimuli, because wolves obviously look the same both to sheep and to other members of their own species. That is to say, if a wolf sees a wolf, and if a sheep sees a wolf there seems to be no difference with regard to how the sensory raw sensory material appears. But there is an obvious difference as to what meaning is given to this raw material. While the sheep judges a wolf to be dangerous, a wolf will judge another wolf to be friendly.

Still, two objections could be raised against the argument that estimative judgments are rightly called judgments. First, one could object that the argument fails to give reasons for why it is not only the estimative power but also other internal, as well as external senses that are said to judge despite the fact that the latter, in particular, do not deliver anything but the raw material mentioned above. Second, one might argue that the term 'judgment' is simply applied here in a very broad sense. One could even say that it is only employed by analogy. And so to ascribe the capacity of judging to nonhuman animals means to speak metaphorically, if not anthropomorphically. However, what lurks behind each of these objections is the view that judgments are propositions. As mentioned in the introductory section, this view is not originally a modern position but was also prevalent during the medieval period. Nevertheless, as

9 See ibid., c. 3, ed. van Riet (1968), 35f.: "Dicemus ergo quia aestimatio excellentior iudex est in animalibus, quae iudicat ad modum adinventae imaginationis cum non est certa, et hoc est sicut id quod accidit homini cum putat mel sordidum quia simile est stercori: aestimatio enim iudicat ita esse et anima sequitur ipsam aestimationem, quamvis intellectus improbet."

10 This is also how Tellkamp (2016), 116, describes the function of the inner senses.

we will see in the following chapters, various medieval authors did not refrain from speaking of judgments when talking about sensory operations. This is not to say, though, that they were uncritical of the ambiguity of this terminology. But instead of abandoning it, many of them resorted to scrutinising the various differences that exist between sensory and intellectual judgments.

CHAPTER 15

Natural Judgments (Thomas Aquinas)

The ascription of judging to nonhuman animals gains its plausibility from the idea of sensory judgments that has been sketched above. The question is, however, what distinguishes human from nonhuman forms of judging apart from the fact that only human judgments are predicative judgments? According to Aquinas, the kind of judgment one finds in nonhuman animals is a 'natural judgment' (*iudicium naturale*).[1] But what does it mean to say that the sheep delivers a *natural* judgment when running away from a wolf? Does it mean that the sheep's reaction of flight is just as natural a process as, for instance, the falling of a stone? No, Aquinas says. Because a stone is a non-living thing, and hence lacks any kind of cognitive powers. Its falling downwards is not triggered by any cognitive process, let alone by some kind of judgment.[2] Nevertheless, there is some similarity between a stone and a sheep. As the stone cannot *not* fall down to earth, the sheep cannot *not* run away when seeing a wolf, as we have seen. This is why Aquinas emphasises that the sheep's judgment is not a *free* judgment. And this is actually one of the main differences between the judgments of humans and those of nonhuman animals. But what does this difference amount to concretely?

As we have seen, the sheep's judgment about the wolf's hostility derives 'from natural estimation' (*ex naturali aestimatione*).[3] In other words, the sheep judges 'by natural instinct' (*ex naturali instinctu*).[4] When the sheep sees a wolf it does not pause for a moment and cogitate about what would be the best thing to do. Rather, it immediately judges that it should flee because this is the judgment that has been 'instilled [to it] by nature' (*a natura inditum*), as Aquinas holds.[5] In contrast, because they are endowed with the freedom of choice

1 See, for instance, Thomas Aquinas, *Summa theologiae* Ia, q. 83, a. 1, co., ed. Leonina v (1889), 307: "Iudicat enim ovis videns lupum, eum esse fugiendum, naturali iudicio, et non libero, quia non ex collatione, sed ex naturali instinctu hoc iudicat. Et simile est de quolibet iudicio brutorum animalium."

2 See ibid.: "[…] quaedam agunt absque iudicio, sicut lapis movetur deorsum; et similiter omnia cognitione carentia. Quaedam autem agunt iudicio, sed non libero; sicut animalia bruta."

3 Thomas Aquinas, *Quaestiones disputatae De veritate*, q. 24, a. 2, ed. Leonina XXII.3.1 (1973), 686: "Sed hoc iudicium est eis ex naturali aestimatione, non ex aliqua collatione […]."

4 See n1 above.

5 Thomas Aquinas, *Summa theologiae* Ia, q. 59, a. 3, co., ed. Leonina v (1889), 95: "Quaedam vero agunt quodam arbitrio, sed non libero, sicut animalia irrationalia, ovis enim fugit lupum

 | DOI 10.1163/9789004363779_020

(*liberum arbitrium*) humans are capable of making a decision about what they should do before doing something. Prior to acting we can measure one option against another. We arrive at our judgments 'by comparison' (*ex collatione*), and so our judgments are not "'one-way judgments' which lead to 'one-way actions'," as Dominik Perler put it.[6] Rather, we are free to arrive at very different judgments about one and the same thing. Such differences are possible not only between members of the human species but within the lives of individuals. Some people find roses beautiful, while others find them disgusting. And similarly one and the same individual might judge that roses are beautiful at one time but find them less beautiful at another time. Nonhuman animals, by contrast, do not have the choice between different judgments about one and the same thing. All members of one species judge in the same way about like stimuli. In other words, they judge as they have been 'programmed' to judge.[7]

This point is also stressed by those interpreters of Aquinas who describe the natural judgments of nonhuman animals as "associative judgments."[8] 'Associative' here does not mean that a sheep freely associates with the wolf whatever it likes to associate with it. It does not associate harmfulness with one wolf, but harmlessness with another. Instead, its judgment is associative in the sense that seeing a grey, furry thing is automatically (or instinctually) associated with perceiving harmfulness. Technically speaking, the perception of sensible forms such as colours, shapes, and so forth is linked to the estimative perception of certain intentions. These intentions do differ, of course. Not all sensible forms are linked to the same intentions. The form of a wolf is linked to different intentions than the form of a blade of grass. Moreover, intentions are relational features in the sense that wolves are not perceived as harmful by all animal species. While sheep are scared of wolves, lions are not. But depending on the nature (or genetic 'programme') of its species, an animal perceives a sensible form as associated with a particular intention. It does not freely form a predicative judgment about this or that thing in the way we would.[9]

However, one could raise two questions here. First, is it really correct to say that humans are free to judge? For would we not also instinctually run away

---

ex quodam iudicio, quo existimat eum sibi noxium; sed hoc iudicium non est sibi liberum, sed a natura inditum."

6 Perler (2006), 87.

7 I borrow this idea from Pasnau (2002), 269.

8 Perler (2012b), 37; Davids (2017), 196.

9 This is also stressed by Frede (2001), 170, who says that estimation "works purely by instinct, i.e., the animal associates with a sensory impression the feeling that something is detrimental."

when encountering a wolf? Second, is it true that nonhuman animals are not free at all? After all, if they are not, what is the difference between sheep and stones, apart from cognition? To the first question Aquinas would possibly reply that humans' judgments do indeed resemble nonhuman animals' judgments in various situations. Although we do not have an estimative power, in his opinion, we do, nevertheless, sometimes react in an instinctual manner. This is particularly obvious in the behaviour of human infants, as Aquinas observes. A child usually takes its mother's breast without any kind of teaching. It does this instinctually.[10] Similar instinctual behaviours might be found in adults, too. Still, what makes the difference between our instincts and those of a sheep is that we are capable of correcting our judgments by virtue of our intellect. So if we judge, for instance, that we should flee from a wolf, we could correct this judgment if someone told us that wolves are much more afraid of us than we are of them. Or, if we saw a wolf in a zoo our intellect would tell us that we need not be afraid of it because it cannot escape its enclosure and hence can do us no harm. In other words, we are in a position to "penetrate" our emotions by our intellectual judgments.[11] Thus, for any kind of judgment we deliver there exists at least the possibility of being corrected.

This point is also important with respect to the question of error. While we are usually capable of realising that we have delivered an erroneous judgment, nonhuman animals lack this capacity, according to Aquinas' theory. 'They do not judge about their own judgment' (*nec iudicant de suo iudicio*), as he expresses it.[12] In modern terms one could say, they do not have *second-order judgments*.[13] This implies two things: first, they do not judge about their judgments in the sense that they deliver a judgment like 'This is a judgment'. In other words, they do not know that they are actually judging because they lack a meta-cognitive level. This is not to say that they lack all awareness of their 'inner life', so to speak. By virtue of their common sense they do, for instance, register whether or not they are seeing at the moment. But this is possible only because their common sense is located on a higher level than their external

10 See Thomas Aquinas, *Commentum in II Sententiarum*, dist. 20, q. 2, a. 2, ed. Mandonnet (1929), 515: "Ad quintum dicendum, quod alia animalia non prosequuntur conveniens et fugiunt nocivum per rationis deliberationem, sed per naturalem instinctum aestimativae virtutis: et talis naturalis instinctus est etiam in pueris; unde etiam mamillas accipiunt, et alia eis convenientia, etiam sine hoc quod ab aliis doceantur." On this passage see also Klubertanz (1952), 159, and Tellkamp (2013), 629f.

11 See Perler (2012b), 33f., as well as id. (2006a), 86f.

12 Thomas Aquinas, *Quaestiones disputatae De veritate*, q. 24, a. 1, ed. Leonina XXII.3.1 (1973), 686.

13 This is also how Davids (2017), 196f., reads this passage.

senses. In the case of estimative judgments, however, there is no higher power that allows them to judge about the judgments of their estimation. This also entails a second point, namely, that they do not judge about their judgments in the sense of evaluating their own judgments. They cannot step back and judge whether the judgment they have delivered previously is right or wrong. In modern terms one could say that they do not grasp the truth-conditions of their judgments. Consequently, they cannot correct their own judgments.

One might doubt whether nonhuman animals' judgments are incorrigible, because if one takes the example of the dog that learns to fear sticks, one could argue that the dog can also be trained *not* to fear sticks anymore. And so the dog seems to be capable of *correcting* its previous judgment about the stick's painfulness. Aquinas does not explicitly consider this possibility, but he is clear that many animals can be trained to react to a certain stimulus in a certain way. So a human trainer can change the way in which an animal naturally judges concerning certain stimuli.[14] Of course, Aquinas does not say that all animals can be trained. For instance, those animals that lack the sense of hearing, such as bees, are immune to teaching because they cannot understand the signs of a trainer.[15] Moreover, it seems unlikely that sheep can be taught not to fear wolves anymore although it is theoretically possible that one could teach a sheep to suppress its natural impulse of flight when seeing a tame wolf, for example. However, teaching is not based on the correction of a first-order judgment by a second-order judgment. It is not that a dog learns to judge that its previous judgment about a stick was an incorrect judgment. Rather, its previous judgment about the stick, say, 'This is painful', is substituted by another judgment about the stick such as 'This is harmless'. Such a substitution of

14 See Thomas Aquinas, *Quaestiones disputatae De veritate*, q. 24, a. 2, ad 7, ed. Leonina XXII.3.1 (1973), 687: "[…] sed in brutis est iudicium naturale determinatum ad hoc quod id quod uno modo proponitur vel occurrit, eodem modo accipiatur vel fugiatur. Contingit autem ex memoria praeteritorum beneficiorum vel flagellorum ut bruta aliquid apprehendant quasi amicum, et prosequendum vel sperandum; et aliquid quasi inimicum, et fugiendum vel timendum: et ideo post flagella, ex passione timoris, quae inde eis insurgit, inducuntur ad obediendum nutui instructoris."

15 See Thomas Aquinas, *In duodecim libros Metaphysicorum Aristotelis expositio*, lib. 1, lec. 1, n. 12, eds. Cathala & Spiazzi (1964), 8: "Inter ea vero, quae memoriam habent, quaedam habent auditum et quaedam non. Quaecumque autem auditum non habent, ut apes, vel si quod aliud huiusmodi animal est, licet prudentiam habere possint, non tamen sunt disciplinabilia, ut scilicet per alterius instructionem possint assuescere ad aliquid faciendum vel vitandum: huiusmodi enim instructio praecipue recipitur per auditum: unde dicitur in libro *De sensu et sensato*, quod auditus est sensus disciplinae." Apparently, Aquinas does not consider the possibility of teaching via other senses such as sight.

judgments leads to a change in the dog's reaction, and this openness to change is what sets cognitive agents apart from non-cognitive beings such as stones. Stones will always fall to earth unless there is some external hindrance. We cannot teach a stone to not fall. Animals, in contrast, are at least potentially capable of changing their reactions, even though this will require some very intense training and might not always be successful.[16]

Furthermore, according to Aquinas, nonhuman animals are superior to non-cognitive things as they even have 'some sort of choice' (*quodam arbitrium*) or 'some likeness of free choice' (*quaedam similitudo liberi arbitrii*).[17] But how is this choice to be understood? Briefly speaking, he thinks that nonhuman animals possess some likeness of free choice because all of their actions need to be preceded by a judgment. That is to say, they will not do anything, unless they judge that they should do it. However, one needs to be careful here, because Aquinas is not suggesting that they are free to choose whatever seems to be the best option, as has been shown. Rather, they are free to the extent that they show what he calls an 'indifference towards acting and not acting' (*indifferentia ad agere et non agere*).[18] That is, they are not determined to act if they do not judge at all.[19] Again, this is why they are quite free in comparison to stones, for example. But in comparison to human beings, they possess only a likeness of freedom because 'their judgment is naturally determined to one

16 For an overview of thirteenth-century views of animal learning see Köhler (2014), 230–240.

17 See Thomas Aquinas, *Summa theologiae* Ia, q. 59, a. 3, co., ed. Leonina v (1889), 95, as well as *Quaestiones disputatae De veritate*, q. 24, a. 2, ed. Leonina XXII.3.1 (1973), 686: "Et similiter est in eis quaedam similitudo liberi arbitrii, in quantum possunt agere vel non agere unum et idem, secundum suum iudicium, ut sic sit in eis quasi quaedam conditionata libertas: possunt enim agere, si iudicant esse agendum, vel non agere, si non iudicant. Sed quia iudicium eorum est determinatum ad unum, per consequens et appetitus et actio ad unum determinatur [...]." See also ibid., q. 23, a. 1, ed. Leonina (1973), 653.

18 Thomas Aquinas, *Quaestiones disputatae De veritate*, q. 24, a. 2, ad 3, ed. Leonina XXII.3.1 (1973), 686: "[...] quamvis sit indifferentia ad agere et non agere in bruto, considerata ipsa actione secundum seipsam, tamen considerato ordine eius ad iudicium, a quo provenit quod est determinatum ad unum, etiam ad ipsas actiones obligatio quaedam derivatur ut non possit in eis inveniri ratio libertatis absolute. Tamen, dato quod in eis esset libertas aliqua, et iudicium aliquod, non tamen sequeretur quod esset in eis libertas iudicii, cum iudicium eorum sit naturaliter determinatum ad unum."

19 Therefore, the translation by Ramelow (2013), 211, is misleading because it says that nonhuman animals "can act if they judge that they should act or they cannot act if they judge against it." [my translation from the German; A.O.]. According to the Latin, however, they do not have the capacity to judge either *for* or *against* something, as Ramelow insinuates. Rather, what Aquinas wants to say here is that they need not act unless they judge *that* they should act. However, they cannot judge on *how* to act.

thing' (*iudicium eorum sit naturaliter determinatum ad unum*), as Aquinas puts it. Unlike us, they do not have the capacity to deliberate about what would be the best thing to do. Even in the above-mentioned cases in which they have learned to react in a way that differs from their natural instincts, they are determined to have only *one* reaction. Once the dog has been taught to fear sticks, it cannot *not* fear sticks.

But what about a situation in which an animal is confronted with, for instance, a piece of food and an enemy at the same time? Would it not be forced to deliberate about what is more important – eating or fleeing? Although Aquinas does not discuss such a case, it is rather unlikely that he would grant an animal the capacity of deliberation in this situation, because the comparison of different options would require the cogitative power at least, if not intellect and reason.[20] Aquinas, however, grants nonhuman animals neither of these faculties. Therefore, he would presumably argue that the animal behaves according to the judgment that is *naturally* stronger in this situation, namely, the judgment that it needs to flee the enemy. This implies that there is some sort of 'hierarchy' of natural judgments, according to which fleeing enemies is prioritised above eating food, for example. But even from a modern perspective, there is nothing fundamentally wrong with such a hierarchy because although most people nowadays would not claim that God has created this hierarchy, they would agree that, depending on the situation, some instincts are stronger than others. Thus, from the fact that an animal favours fleeing over eating one cannot infer that it has judged that fleeing is more important than eating. Rather, it follows a natural impulse.

Still, it is misleading to say that Aquinas takes nonhuman animals to be subject to "total determinism," as Stephen Loughlin suggests, because *total* determinism is rather what applies to non-cognitive things, such as stones.[21] As has been said, the stone's falling down is not the result of an act of judgment, whereas nonhuman animals' actions are based on judgments. Therefore, animals can be said to possess at least 'some freedom' (*aliqua libertas*), in the view of Aquinas.[22] To be sure, their freedom is a *likeness* of freedom rather than freedom proper. Nevertheless, they are not merely subject to the laws of nature as stones are. In short, their capacity of delivering natural judgments does not lift their cognition to the level of human cognition, but it enables them to interact with different things in ways which are not open to non-cognitive beings, and in this sense they are similar to us.

---

20 For a different reading of Aquinas see Rubini (2015), 472.

21 Loughlin (2001), 49.

22 See n18 above.

CHAPTER 16

Erroneous Judgments and Differences in Estimation (Albertus Magnus)

As we have seen, Thomas Aquinas' theory of natural judgments relies heavily on the Aristotelian idea of sensory judgments, as well as on Avicenna's theory of estimation. The same could be said about Albert's account of judging in nonhuman animals because he also thinks that they engage in judging in the sense that they interact with the world by virtue of their external and internal senses, most importantly, by virtue of estimation. It is, nevertheless, worth looking at Albert's account for a moment, because in some places he focuses on different aspects than Aquinas does. The first interesting aspect is the possibility of error.[1] Of course, this aspect is not totally absent from Aquinas' theory, as we have seen in the previous chapter. But Albert's statements on error in nonhuman animals shed more light on the nature of their judgments. The second interesting aspect concerns the question of whether the capacity of judging varies among different species of animals.

As regards the aspect of error, Albert provides a strange but illuminating example. In his *Quaestiones super De animalibus*, he discusses why wolves are less timid in the field than in the forest. His answer is as short as it is simple: in the forest, wolves often mistake trees for human beings, therefore, they often get frightened. In the field, by contrast, they cannot fall prey to this error, which is why they are much braver there.[2] This might seem like a strange example, because wolves usually live in the woods. However, for the purpose of the present chapter the consistency of this example is not what matters. Rather, what matters is that wolves are said to err because they mistake trees for human beings.[3] More precisely, they 'estimate' (*aestimant*) that what they see is a

1 On this question see also Oelze (forthcoming b).

2 Albertus Magnus, *Quaestiones super De animalibus*, lib. VIII, q. 32, ed. Filthaut (1995), 199: "Ad quartum dicendum, quod quia lupus est animal timidum, ideo frequenter aestimat arbores esse homines, sed cum est in campo, tunc melius potest undique respicere et sibi praecavere et ideo audacior est extra silvam quam in silva." On wolves in Albertus Magnus' works see Anzulewicz (2013b).

3 There are a number of similar examples in medieval texts. For instance, in his treatise on the soul, an anonymous master of arts states that donkeys are much more afraid of water than of lions since they become frightened when seeing their own reflection in the water, see Anonymous, *De anima et de potenciis eius*, ed. Gauthier (1982), 46. Similarly, a medieval

 | DOI 10.1163/9789004363779_021

human being and not a tree. Albert, of course, employs this term because the wolf's erroneous judgment is brought about by its power of estimation in this situation. But if one leaves aside the particularly medieval theory of estimation for a moment, the example turns out to be very similar to the example famously used by Norman Malcolm and Donald Davidson in their discussion on whether nonhuman animals can think.

Their example is the following:[4] a dog chases a cat through a garden. The cat heads towards an oak tree but, at the very last moment, jumps onto a nearby maple tree. The dog misses this manoeuvre and so it stops at the oak tree and begins to bark up the wrong tree. In order to explain the dog's behaviour, most people would probably say that the dog erroneously *thinks* or *believes that* the cat climbed up the oak tree. For why should it bark up the oak tree if it did not think or believe that the cat was sitting up there? Now, the way in which we are inclined to express the dog's belief or thought is a typical propositional judgment of the form 'X believes/thinks that F'. According to Davidson, we thus ascribe what he calls "propositional attitudes" to the dog.[5] But Davidson is skeptical about whether we are right to attribute such attitudes to nonhuman animals. In short, his "lingualist master-argument," as Hans-Johann Glock calls it, runs as follows:[6] thought requires concepts and concepts require language. Consequently, if a creature lacks the capacity for language it lacks both concepts and any kind of thoughts, no matter whether those thoughts are desires, beliefs, hopes, or anything else of this sort.

Davidson's argument for the language-dependence of thought is, of course, much more intricate, and it has given rise to various objections. But suffice it to say that it also has consequences for the 'capacity' of erring. Since the dog has no propositional attitudes, it also lacks the capacity to realise that it is mistaken about something. In short, if the dog is incapable of having the belief 'The cat is sitting in this tree' it is likewise incapable of falsifying or confirming this belief. Interestingly, this is exactly what seems to apply to estimative judgments, as well. If the wolf's estimative power judges that this thing over there is a human being and not a tree there is no obvious way to correct this error because, like Avicenna, Albert thinks that the correction of false sensory judgments (e.g. by imagination or estimation) requires the intellect.[7] This is

bestiary tells the story of a dog that mistakes its own reflection for another dog, see Clark (2006), 147.

4 See Malcolm (1972), 13; Davidson (1982), 319.

5 Davidson (1982), 318f.

6 See Glock (2010), 15. See also Wild (2013), 90–104.

7 See, for instance, Albertus Magnus, *De animalibus*, lib. XXI, tr. 1, c. 3, ed. Stadler (1920), 1331: "Sed in omnibus hiis non movetur nisi ex fantasmate: et ideo frequenter errat sicut et alia

quite plausible for the following reason: imagine the wolf's estimation judges that X is a human being at time t_1. Now at t_2 it delivers a different judgment, namely, the judgment that X is a tree. No matter *why* it judges differently about X at t_2 (but let us, for the sake of the argument, suppose that it does), it is incapable of registering *that* it judges differently at t_2 because, like any other sensory power, estimation cannot reflect on its own acts. Therefore, it seems impossible for it to correct its own judgments. Maybe one could say that it can *alter* its judgments, but the only way in which judgments can be *corrected* is by the aid of the intellect.[8]

But what about highly-developed nonhuman animals, such as pygmies, which have what Albert calls 'a shadow of reason'? Are they as prone to error as other animals? In short, the answer is: they are. The possession of a shadow of reason does not prevent them from making mistakes. The main reason for this is that this shadow of reason lacks a connection with the intellect. Nonetheless, pygmies and apes in general usually outdo other species of nonhuman animals in judging whether something is harmful or helpful, according to Albert. In particular, they seem to have developed a sense for the needs of others. That is, they not only deliver judgments about what is good or bad for them but their judgments also extend to what is good or bad for other members of their own species, if not even for the members of other species.

In order to illustrate this, Albert mentions the example of an infant ape that points out its mother's breasts to another young ape.[9] It seems not only to care for its own well-being but also for the well-being of others. From a modern

animalia quia sicut diximus in antehabitis, ubi fantasticum intellectui non coniungitur, frequens incidit error […]."

8 Therefore, I disagree with Tellkamp (2016), 119f., who claims that "[s]ince animals do not possess concepts, Albert does not relate truth and falsehood to propositions, but rather to *intentiones* produced by phantasy." It is true that an animal's estimative perceptions are either true or false but the animal itself does not know what the truth-conditions for its estimative judgments are, as Tellkamp (2013), 207, claims.

9 Albertus Magnus, *De animalibus*, lib. XXI, tr. 1, c. 3, ed. Stadler (1920), 1331: "Videtur autem genus symiarum prae ceteris animalibus brutis aestimare de nocivo et conferente ex sensu et ymaginatione et memoria. Aliis enim aestimationem habentibus de sibi conferentibus et nocentibus, et illa persequentibus vel fugientibus, symia aestimat de conferentibus sibi et aliis animalibus: et ideo videns parvulum exhibet parvulo ubera, non propria, sed matris quae peperit eum si permittatur: puero enim exhibet ubera feminarum si permittatur: et hoc alia non faciunt animalia: et in hoc cognoscitur symia melioris esse aestimationis quam aliud animal. Et haec in exterioribus convenientia etiam interiorem significat convenientiam: quia sicut diximus, aestimativa ipsorum similior est rationi quam aestimativa aliquorum aliorum animalium." According to Janson (1952), 89, the example of an infant ape helping another member of its species is quite original in medieval ape lore.

perspective, one could say that apes thus engage in social cognition.[10] According to some scholars, they even have what is called a 'theory of mind'. This means that they are capable of ascribing mental states, such as intentions and desires, to others. The infant ape could be taken to do this because, by offering its mother's breasts to another member of its species, it somehow shows awareness for and understanding of the latter's needs and desires. Of course, there is a great deal of debate in contemporary research over whether apes have a theory of mind or not.[11] But, for Albert, the behaviours of apes clearly show that there is a great deal of similarity or 'conformity' (*convenientia*) between us and them – not only as regards physiology but also with a view to psychology. Their power of estimation is much more similar to reason than the estimative power of other animals, he claims. Thus even though they are not immune to making mistakes, their estimative judgments are, in his view, superior to the judgments of other animals.

10 See Tomasello & Call (1997), esp. part II.

11 The most important contributions to this debate include Premack & Woodruff (1978); Tomasello & Call (1997), 311–341; Povinelli & Giambrone (1999); Call & Tomasello (2008).

CHAPTER 17

Reflective and Experimental Judgments (Peter of John Olivi, John Buridan)

As seen, both Thomas Aquinas and Albertus Magnus very closely adhere to Avicenna's theory of estimation and the idea of estimative judgments. Both of them attend to a number of aspects concerning estimative judgments, including the question of whether such judgments are entirely instinctual or whether they can be changed, if not corrected. Two other authors, namely, Peter of John Olivi and John Buridan, also address these questions but take them further. Olivi, in particular, wonders whether nonhuman animals are capable of reflecting on what would be the best thing to do in a certain situation. Buridan, for his part, focuses on how they acquire certain judgments over time through learning or experience.

The question of reflection has already been mentioned in connection with Aquinas insofar as we have asked how Aquinas could explain the behaviour of an animal that is confronted with a situation in which it must choose between pursuing a piece of food and fleeing an enemy (see Chapter 15). Aquinas could readily answer this question, we have said, by pointing to some sort of 'natural hierarchy' of estimative judgments. In other words, the animal does not reflect on whether it should eat or flee because reflection usually requires the cogitative power which nonhuman animals lack. Rather, the judgment that it should flee is *naturally* stronger than the impulse to eat, because running away from an enemy is more important for survival than eating is. Of course, without eating it will not be able to flee in the long run. But in the particular situation in which it can only flee or eat, fleeing is to be preferred. The animal itself, however, does not gain this insight. This is because, first, it lacks the appropriate cognitive power, and second, because it is incapable of grasping the reason for which something should be done or not be done.[1]

For Olivi, even the simplest estimative judgments require the 'apprehension' (*apprehensio*) of at least three aspects. For instance, when a sheep judges that this wolf is hostile, it first apprehends what Olivi calls the '*ratio inimicitiae*'. In short, this means that it grasps the feature of hostility. But the grasp of this

1 See, for instance, Thomas Aquinas, *Summa theologiae* Ia, q. 59, a. 3, co., ed. Leonina v (1889), 95, where he explains that nonhuman animals fail to grasp, for instance, the "*ratio universalis*" of goodness. Thus, they do not understand why something is good or bad.

 | DOI 10.1163/9789004363779_022

feature is not sufficient for triggering a reaction of flight, according to Olivi. In his view, the sheep also needs to apprehend that the grey, furry thing it sees is exactly what has this feature. Moreover, third, it also needs to have an understanding of itself as being the 'goal' (*terminus*) of the wolf's hostility. Thus, every judgment about what is to be done in a certain situation also involves an awareness of these three aspects at least.[2] For Olivi, this applies even to relatively simple situations, such as the sheep's encounter with a wolf. In this case, there are not that many options for the sheep: it can either flee or remain. But what about more complex situations, such as the above-mentioned choice between fleeing and eating?

Olivi considers a relatively similar example: a dog or a snake is attacked by another animal. The problem now is that neither the dog nor the snake can protect its whole body against the attacker's bites and blows. Therefore, they have to decide what is the most important part to save: the head or something else? They both protect the head instead of another part of the body because both judge that the head is a more important part of the body than the tail or the rear quarters. For Olivi, this shows that they possess 'some general power' (*aliqua communis potentia*) by means of which they can compare two extreme options with each other.[3] One extreme option consists in protecting no part at all. The other option is to save at least one part or some parts. Of course, if the animal chooses the latter option, other parts of its body might be severely wounded. However, this is still better than being wounded everywhere. So, in

2 See Peter of John Olivi, *Quaestiones in secundum librum Sententiarum*, q. 64, ed. Jansen (1924), 603: "Quando enim ovis aestimat lupum sibi esse inimicum, oportet quod apprehendat illam rem quam sibi iudicat inimicam; apprehendere enim solam rationem inimicitiae non est apprehendere lupum sibi esse inimicum. Unde etiam ultro duo praedicta oportet quod simul apprehendat se tanquam terminum illius hostilis respectus." On this passage see also Toivanen (2013a), 284, On Olivi's theory of animal actions in particular see Toivanen (2011).

3 Peter of John Olivi, *Quaestiones in secundum librum Sententiarum*, q. 62, ed. Jansen (1924), 587f.: "Ergo sicut illam appetitivam oportet dominari omnibus membris et sensibus quos ad suos actus applicat vel ab eis retrahit: sic oportet unam iudicativam sibi assistere quae de omnibus actibus eorum iudicet et eorum delectationes vel dolores advertat et alteram alteri praeferat vel praeferandum ostentat. Praeterea, quando canis vel serpens pro conservatione capitis exponit aliud membrum aut pro conservatione totius exponit aliquam partem, tunc praefert totum parti et caput alteri membro. Ergo oportet in eis esse aliquam communem potentiam quae in simul ambo extrema et mutuam eorum comparationem et unius ad alterum praeferentiam ostendat, quamvis non cum illa plenitudine et altitudine reflexivi iudicii cum qua fit hoc ab intellectu." On this passage see also Roling (2011), 238, and Toivanen (2013a), 281–283.

short, the animal *reflects* on what is the best thing to be done and it engages in this reflection by virtue of its common sense.[4]

Still, Olivi stresses that nonhuman animals' 'reflective judgment' (*iudicium reflexivum*), as he calls it, does not have the same 'completeness and altitude' (*plenitudo et altitudo*) as the reflective judgments of the human intellect.[5] But how is this to be understood? One possible reading would be that nonhuman animals cannot reflect on just any kind of issue. Rather, their inner sense makes them reflect only on what is immediately relevant for their survival or well-being. Moreover, one could argue that their reflective judgments arise from instinct much more than they arise from deliberation. This is also a possible objection to Olivi's theory, because it is hard to tell from the outside whether a dog's or a snake's reaction in protecting a certain part of the body derives from instinct or from some process of reflection. Even humans, one could say, *instinctually* protect their heads, for example, when attacked by somebody. However, Olivi could reply to this objection by maintaining that this is exactly the reason why he takes those judgments to be inferior to intellectual judgments: they do not arise from a full-fledged process of reflection. Nevertheless, the animals' inner sense 'notifies a preference' (*ostendat praeferentiam*) for a certain way of acting, as he puts it, and this is at least similar to what the intellect does when judging that option A is better than option B.

This might still not prove an entirely satisfying answer because what would be the difference between a natural hierarchy of instinctual judgments and an instinctual preference for a certain reaction? Olivi might say that there is no significant difference. However, in the passage in which he discusses the reflective judgments of nonhuman animals he points out that their judgments about what they should do also depend on whether a certain action would give them pleasure or pain.[6] Naturally, they will try to avoid pains. But they cannot know whether something will give them pleasure or pain unless they have already experienced a similar thing or situation. So if they remember that this or

---

4 According to Olivi, there is a unity, rather than a multiplicity of inner senses. Thus, estimation, for example, is not a distinct inner sense but a particular kind of operation of the common sense. On this theory see Toivanen (2007) and (2013a), 247–265. Olivi shares this position with Buridan, see De Boer (2014) and Lagerlund (2017).

5 In another passage he adds that nonhuman animals do not judge "with full affirmation" (*plena affirmatione*), see *Quaestiones in secundum librum Sententiarum*, q. 57, ed. Jansen (1924), 327.

6 See n3 above. To what extent this might involve self-awareness is discussed by Toivanen (2013b), 375.

that action has given them pain before, they are likely to judge that they should refrain from doing it. This judgment, however, arises from experience rather than from instinct. But do nonhuman animals judge by experience at all?

While Olivi does not explicitly discuss this question, John Buridan addresses it, as we have already seen in Chapter 12. In short, his view is that delivering an 'experimental judgment' (*iudicium experimentale*), as he calls it, is not a particularly human capacity. But what exactly is an experimental judgment? Buridan gives the following example:[7] if you see a fire and touch it or come close to it, you will perceive that the fire is hot. The same will happen with a second fire and, maybe, a third and so forth. But after a while you no longer need to touch fire in order to find out that it is hot. Rather, you will judge *by experience* that it is hot.[8] But why is this not simply a judgment by memory, one could ask, since my judgment that some fire C is hot derives from my *memory* of fires A and B, and so there seem to be good reasons to believe that all we need is memory.

To be clear, Buridan does not deny that memory plays a role here. However, like many other thinkers (as we will see more clearly in Part 5) he has a relatively narrow definition of memory. Strictly speaking, memory is nothing but the recognition of something. That is to say, if one sees fire A one can *remember* this fire to be hot only if one has seen (and touched) *this particular* fire A before. By contrast, if one judges by experience one judges with respect to a given fire C that one has *not* seen before. Admittedly, experience is based on memory to some extent because the experimental judgment that fire C is hot relies on the memory of the encounters with fires A or B. But there is an additional element involved, namely, the capacity to cognise similarities which has already been mentioned in Chapter 12. When seeing a given fire C one cannot, strictly speaking, remember that it is hot because this supposes that one has seen fire C before. Instead, one realises that, for instance, the shape, the colour, and the smell of C is *similar* to the shape, the colour, and the smell of A and B. This then provides the basis for applying the memory of A and B to C. And so

7 This example has already been discussed in Chapter 12. For a discussion see also Pluta (2015), 276f.

8 See John Buridan, *Quaestiones in Analytica Priora*, lib. II, q. 20a, ed. Hubien [undated typescript]: "Verbi gratia, si tu ad sensum cognovisti quod ignis A erat calidus, et postea idem de igne B, et sic de multis aliis, tu postea videns ignem C, et non tangens ipsum, judicabis per memoriam de aliis et propter similitudinem quod ille ignis C est calidus; et hoc non est, proprie loquendo, judicium per sensum, quia non tangis ipsum, nec solum per memoriam, quia memoria proprie non est nisi prius cognitorum et tamen ipsum ignem C numquam alias vidisti nec cognovisti; sed hoc judicium vocatur 'experimentale'."

one judges by experience that fire C is hot even though one has neither seen nor touched this particular fire before.[9]

This form of judgment is not an exclusively human capacity, according to Buridan. In his view, other animals are capable of judging by experience, as well. To return to a previous example, dogs that begin to fear sticks or stones after they have been beaten with them detect the similarities (shape, colour, etc.) between stone C and stones A or B. And by remembering the painfulness of A and B they judge that C is likely going to be painful, too.[10] Now one might ask whether this requires the formation of the concept 'stone' because it seems an experimental judgment amounts to something like 'Stones are hurtful'. However, as explained in Chapter 12, one needs to be careful here. Because even though it seems as if the dog delivers a *universal* judgment about the painfulness of stones, Buridan is keen to emphasise that the judgment is *particular*, not universal because the dog does not come to its judgment about the painfulness of a stone C by *abstracting* from its encounters with stones A and B.[11] Rather, it spots the similarities between A, B, and C and then applies its memory of A and B to C.

Now, one could argue that spotting similarities is or involves some sort of abstraction because A is not identical with B or C. But for Buridan (as for the other thinkers covered in this study), the cognition of similarities does not necessarily involve the formation of concepts or the cognition of universals, as seen in Part 2. Of course, in humans it does (or might) lead to concept formation and universal cognition, at least if one holds that intellectual abstraction is based on sensory cognition. But this is not the case in nonhuman animals because by definition they lack intellectual powers and thus only judge by their senses.[12]

---

9 Someone like Aquinas could not readily account for such a cognition because "the recognition of a commonality as shared by several individuals" requires the cogitative power, as Barker (2012b), 62, notes.

10 See John Buridan, *Quaestiones in Analytica Priora*, lib. II, q. 20a, ed. Hubien [undated typescript]: "Et non solum homines, immo aequaliter brutae hujus modi judicio utuntur; unde propter hoc canis timet lapidem si aliquis laesit ipsum."

11 Ibid.: "Et omnia praedicta principia sunt singularia, et sunt principia in arte vel in prudentia, et non in scientia speculativa vel demonstrativa." See also Klein (2016), 178, and Pluta (2015), 276.

12 See John Buridan, *Summulae de demonstrationibus* 8.5.4, ed. de Rijk (2001), 128f.: "Et non fit hoc per intellectum, quia sic faceret iuvenis catulus vel caniculus, licet non habens intellectum; sed hoc iudicat virtus aestimativa."

CHAPTER 18

The Ascription of Judgments and the Problem of Anthropomorphism (William of Ockham, Adam Wodeham, Gregory of Rimini)

The thinkers discussed in the previous chapters all accepted the following assumption: one can ascribe judgments to nonhuman animals in the sense in which one can ascribe judgments to sensory powers. To be clear, they did not hold that there is no difference between sensory and intellectual judgments. As we have seen, they all emphasised that the sensory judgments of nonhuman animals are in many ways inferior to the intellectual judgments of human beings, mostly because they are instinctual judgments. And even if some degree of learning, reflection, or experience might be involved, as some maintained, they stressed that his does not reach the level of intellectual learning, reflection, or experience. Despite these differences, they held that it makes sense to speak of judgments in nonhuman animals. Thus, they obviously employed a relatively broad concept of judging because they took it for granted that acts of judgment can occur at the level of both the intellectual and the sensory powers. However, this relatively broad concept of judging was discussed critically by authors of the fourteenth century. In this last chapter, we shall look at three of them, namely, William of Ockham, Adam Wodeham, and Gregory of Rimini.

To begin with Ockham, there is an interesting passage in his *Sentences* commentary in which he expresses surprise at the idea that sensory powers engage in judging. Judging, he says, is 'a determinate complex act' (*actus complexus terminative*). That is to say, it is a cognitive operation 'that presupposes the apprehension or formation of something complex' (*praesupponit apprehensionem sive formationem complexi*). Apparently, what Ockham has in mind here is the above-mentioned 'building-block model'. According to this model, judgments are propositions and propositions are built of concepts. In order to judge 'that X is F', one needs to possess the concepts 'X' and 'F', and one needs to know how to arrange them in a certain order. In brief, this is what is meant by the phrase that judging is a *complex* act. And so if one attributes the ability of judging to sensory powers, one grants them 'complex cognition' (*cognitio complexa*).[1]

1 William of Ockham, *Quaestiones in librum tertium Sententiarum* (*Reportatio*), q. 2, eds. Kelley & Etzkorn (1982), 85: "Mirabile est quomodo sensus potest iudicare, cum iudicare sit actus

 | DOI 10.1163/9789004363779_023

But Ockham also thinks that nonhuman animals, human children, and mentally disabled people engage in judging because they differentiate between useful and harmful things.[2] So even those creatures which do not (yet) have intellectual powers or whose intellectual powers are seriously impaired have the capacity to cognise whether something is good or bad for them. Shall we therefore grant them complex cognition? No, Ockham says, because the way in which they judge is very different from the way in which fully rational human beings judge.[3] In his opinion, the difference is the following: in a healthy human adult the act of judging is preceded by an act of apprehension, more precisely, by a complex apprehension. Only then comes the judgment which is brought about by the intellect. In other words, humans usually perceive something at first before they judge about what they perceive or have perceived. Thus, perception and judgment are distinct cognitive acts. They are not only numerically distinct but they are also distinct in that different cognitive powers are at work, namely, the senses in the first instance, and the intellect in the second.

In nonhuman animals, human children, or mentally disabled people, the situation is different. When they judge they cannot judge in addition to or independently from the perceptions they have because this would require (intact) intellectual faculties. Since they lack those faculties (or since those faculties are not properly working as in the case of the mentally disabled) their act of judging is simply the same as the act of apprehension, as Ockham says. And the apprehension they have is only a 'non-complex apprehensive act' (*actus apprehensivus incomplexus*).[4] So, in Ockham's view, the ascription of judgments to such beings does not necessitate the ascription of complex cognition. As long as one makes clear that judging is not a separate, complex intellectual act here, there is no problem with this ascription. However, one might criticise Ockham's argument by saying that he introduces a sort of double standard. On the one hand, he employs a relatively narrow concept of judgments as complex cognitive acts. On the other hand, he calls non-complex apprehensions judgments, too. Although Ockham himself does not address this objection, it

complexus terminative et praesupponit apprehensionem sive formationem complexi, quod non potest fieri per potentiam sensitivam. Unde ponendo iudicium in sensu, oportet ponere quod sensus habeat cognitionem complexam formaliter vel aequivalenter […]."

2 William of Ockham, *Quaestiones in librum quartum Sententiarum (Reportatio)*, q. 14, eds. Wood & Gál (1984), 314: "Ad aliud dico quod sensitiva habet iudicium. Patet in brutis, pueris, fatuis etc., qui iudicant inter nociva et convenientia, sed non habent actum iudicandi distinctum ab actibus apprehensivis incomplexis, sicut intellectus habet."

3 For a brief discussion of those differences see also Perler (2008), 158f.

4 See n2 above.

was exactly this point that gave rise to the famous debate between his secretary Adam Wodeham and Gregory of Rimini.

The debate between Adam Wodeham and Gregory of Rimini emerged in a particular context, namely, the question of how to explain sensory illusions.[5] The main problem can be illustrated by the following example: if we stand aboard a ship we might take the trees on the nearby coast to be moving. But, in fact, it is not the trees that are moving but the ship. Since we do not immediately notice the latter, we are, at least for a while, mistaken about the former. The crucial question is where our misperception actually comes from. Should we blame our sense of sight or our intellect? In Wodeham's opinion, it is the intellect that errs in this case because the intellect delivers an 'incorrect judgment' (*iudicium erroneum*) about the moving trees.[6] That is, strictly speaking, our impression that the trees are moving is not a *sensory* illusion because it originates from an *intellectual* error. However, there is a strong objection that can be raised against this view, as Wodeham concedes. Imagine there is a dog standing aboard the ship, too. Would we not say that the dog also falls prey to the illusion that the trees are moving? Maybe the dog would even be terrified by the moving trees and run away, Wodeham says. This seems like very strong evidence for the fact that the perception or 'apparition' (*apparitio*) of the moving trees is not an intellectual but a 'sensory act' (*actus sensitivus*) because the dog has no intellect, according to the common view.[7]

Wodeham gives the following reply. On the one hand, he notes, there are good reasons to believe that the dog has 'simple visual perceptions' (*visiones simplices*). It is endowed with a visual apparatus by which it can see most of the things we see. So if we see the trees on the nearby coast, a dog standing next to us is very likely going to see these trees, as well. But does the dog make judgments about the trees like we do? Ockham's suggestion was to say that nonhuman animals judge insofar as their judgments coincide with their perceptions. Consequently, the dog's judgment about the moving trees is not a

---

5 This debate has already been reconstructed and analysed by Reina (1986); Tachau (1993), 665; Michon (2001), 326f.; Perler (2006), 89–94. See also Perler (2014a) who reconstructs Ockham's position.

6 See Adam Wodeham, *Lectura secunda in primum librum Sententiarum*, prol., q. 4, §7, ed. Wood (1990), 97: "Sed illa apparitio non est visio sed iudicium erroneum causatum mediante visione."

7 Ibid., §8, 98: "Contra: illa apparitio qua apparet homini quod arbores moveantur est actus sensitivus. Sed, per te, nullum iudicium quo assentitur aliquid esse aliquale, vel aliqualiter se habere, est actus sensitivus. Igitur. Maior probatur, quia etiam brutis apparent arbores illae moveri, in tantum si navis moveretur versus arbores, fugerent [ac] si essent eis terribilia quaedam."

cognitive act that is distinct from its act of seeing, according to Ockham. However, as has been said, one might criticise this suggestion since it introduces a broader concept of judging. But could we also ascribe judgments to dogs and other nonhuman animals in terms of complex cognitive acts?

According to Wodeham, there are at least two problems with such an ascription. The first could be called the *epistemological problem*. This problem arises from the fact that all we can observe is the dog's behaviour. That is, we can watch the dog sitting aboard the ship, looking at the trees and, maybe, running away. However, we cannot look into the dog's head, so to speak. Thus, apart from the fact *that* the dog sees something (and even this might be taken to be uncertain to some degree), we cannot really know *what* the dog actually sees, nor can we really know that its reaction of flight is based on a 'complex objective judgment' (*iudicium complexe obiective*). In short, we can only read the dog's behaviour or the 'effects and movements' (*effectus et motus*), as Wodeham puts it, but we cannot read the cognitive processes or the mental states that are (or might be) the causes of its behaviour.[8] And so if we ascribe the same mental states or cognitive processes to the dog which we take to be responsible for *our* behaviour, we might fall prey to what is usually called *anthropomorphism*. In short, from human-like behaviour we infer human-like cognition.

Anthropomorphism is not necessarily wrong, but it definitely leads to a second problem. This could be called the *metaphysical problem* or *problem of classification*. The problem here is that if we do decide to ascribe judgments in terms of complex cognitive acts to the dog, we are obliged to attribute 'practical reason' (*ratio practica*) to it too, according to Wodeham. Since only rational animals possess practical reason, the dog would then qualify as a 'rational animal' (*animal rationale*). Hence the ascription of judgments would shift the common metaphysical boundary between human and nonhuman animals because dogs (and, maybe, other nonhuman animals, too) would qualify as *rational* animals.[9]

Of course, one could try to cope with both problems. But if one considers the epistemological problem to be unsolvable or the changes in the classification of animals to be unacceptable, one needs to give an alternative explanation

---

8 See ibid., 99: "[…] concedo quod in brutis sunt visiones tales simplices, ad quas natum esset sequi iudicium illud quo complexe [formato] obiective ipsis competeret. Sed utrum eis conveniat, scire non possumus, nisi coniciendo ex effectibus et motibus sequentibus tales visiones simplices."

9 See ibid.: "Sed si hoc movere deberet, esset consequenter in eis ponenda ratio practica, id est dictamen practicum de eligendis et respuendis, prosequendis et fugiendi. Sic enim agunt si dictamen haberent. Et tunc non video quare non debeant animalia rationalia appellari."

for the dog's behaviour which coherently accounts for the dog's movements without referring to judgments.[10] Yet, what would such an explanation look like? For Wodeham, such an explanation would centre on what he calls 'simple apprehensions' (*apprehensiones simplices*). The dog perceives, for instance, the trees on the shore and 'immediately' (*statim*) reacts to this perception (or apprehension). So it does not perceive the trees at first and then takes time to deliberate about whether what it sees is harmful or harmless. Rather, its perception immediately triggers a reaction (e.g. of flight), and this reaction comes 'from natural instinct' (*ex instinctu naturae*), according to Wodeham.[11]

Two things are important to note here. First, like Ockham, Wodeham claims that there is no *separate* act of judging in other animals. Even if one said that the dog *errs* about the movement of the trees like we do, we are still better off than the dog because we can *correct* our error. While the dog runs away frightened, we can reflect on whether what we perceive (or seem to perceive) makes sense or not because our perceptions and our judgments do not (necessarily) coincide. Yet, unlike Ockham, Wodeham does not think that judging and perceiving coincide in nonhuman animals. In his opinion, simple apprehensions should not qualify as acts of judgments at all. If one does call them judgments, nonetheless, one applies the term in an equivocal sense, he says.[12] This terminological rigour is undoubtedly one of the strengths of Wodeham's theory because it avoids an anthropomorphic way of speaking, if not an anthropomorphic fallacy, because we cannot infer that the same cognitive

10 According to Perler (2006), 91f., Wodeham's search for an alternative explanation that avoids the ascription of judgment to nonhuman animals is also motivated by the fact that he holds a theory according to which the content of judgments is a so-called '*complexe significabile*' which is "expressed by a full-fledged sentence." However, this cannot fully explain Wodeham's animosity because Gregory of Rimini also subscribes to this theory. Nevertheless, he grants nonhuman animals a form of judgment that has no *complexe significabile* as its content. On the *complexe significabile* see also Nuchelmans (1973), 227–237, and (1980), 104–106.

11 See Adam Wodeham, *Lectura secunda in primum librum Sententiarum*, prol., q. 4, §8, ed. Wood (1990), 99f.: "Id est non solum non libere prosequuntur ac fugiunt proficua et nociva, sed nec deliberant nec iudicant aliquid iudicio complexo obiective conveniens exsistere vel nocivum de fugiendo vel [proficuum de] prosequendo. Sed statim ad simplicem apprehensionem istius quod est nocivum fugiunt et illius quod est conveniens prosequuntur. Et si talis prosecutio conveniat omnibus individuis alicuius speciei naturaliter, tunc sufficit prima apprehensio simplex ad illam prosecutionem et fugam, quod tantum fit de re praesente. [...] sed hoc est ex instinctu naturae." On the difference between simple and complex apprehensions see also Perler (2006), 90.

12 See ibid., 100: "[...] concedendum est illam apparitionem non esse iudicium nisi aequivocando de iudicio."

processes underlie the dog's behaviour from the simple fact that its behaviour resembles our own. This problem, of course, is well known to modern philosophers. As mentioned in Chapter 12, it has given rise to 'Morgan's Canon', which holds that one should not explain animal behaviour by higher psychological processes (e.g. judging) if it can also be coherently explained by lower psychological processes.[13] It appears that Adam Wodeham anticipates this principle as he says that one can account for the dog's behaviour by reference to simple apprehensions. But is this explanation satisfactory?

For Adam's counterpart, Gregory of Rimini, it is not satisfactory for several reasons. On the one hand, Gregory agrees with Wodeham as far as the epistemological problem is concerned. So he agrees that we do not have immediate access to other animals' mental states, especially if we cannot communicate with them by language. For instance, whether they have 'complex cognition' (*notitia complexa*) is not 'naturally' (*naturaliter*) known to us, as he puts it. All we can do is to observe their behaviour and then try to give an explanation of whether it is 'likely' (*probabiliter*) or not that complex cognition is what has caused this behaviour.[14] On the other hand, however, that is exactly the point where Gregory starts to disagree with Wodeham because in his view Wodeham's theory can only account for those cases of animal behaviour in which an animal always shows the same response to the same stimulus. Maybe the dog aboard the ship is such a case. But what about those cases in which an animal reacts differently to one and the same stimulus?

Gregory gives the example of an animal that sees a piece of bread. According to Wodeham's theory, the animal's reaction to this stimulus would always be the same. But that is not the case, Gregory states because quite often we can see that one and the same animal goes for the bread at this particular time while at a later point it does not. This kind of behaviour cannot be explained with reference to simple apprehensions that immediately lead to instinctual reactions, because instinctual reactions usually remain the same. Instead, one must suppose that every time the animal sees some bread it *judges* whether or not it should eat it. If its judgment at time t_1 is different from the judgment at time t_2 its behaviour will be different, too.[15]

---

13 See p. 92 n10.

14 Gregory of Rimini, *Lectura in primum et secundum Sententiarum*, dist. 1, q. 3, eds. Trapp & Marcolino (1981), 304: "Et hic dico quod, quamvis istud non possit nobis plene naturaliter esse notum, probabiliter tamen ex his, quae percipimus, possumus arguere quod talem notitiam habeant."

15 See ibid.: "Quod primo convinci potest, quia, sicut videmus, aliquando brutam, apprehendit aliquod sensibile, ut puta panem, et movetur ad ipsum; aliquando vero apprehendit

Like the authors we have examined in the previous chapters, Gregory does not claim that this kind of judgment is an intellectual, propositional judgment. Rather, it is a non-propositional 'sensory judgment' (*iudicium sensibile*) or a "low-level judgment," as Dominik Perler put it.[16] Gregory mentions at least two differences between sensory and intellectual judgments.[17] First, sensory judgments do not involve universals but only particulars. Hence, nonhuman animals can only judge about this or that particular bread, but not about bread in general, for example. Second, they cannot judge about 'insensible things' (*insensibilia*), but only about what they perceive by their senses and what has practical relevance for them in this moment. When they perceive a piece of bread they have to judge whether it is something they should eat or not. They cannot, however, judge about the ontological status of bread, its chemical structure, or other theoretical matters. In this sense, their judgements rank lower on the scale of possible acts of judging. And so even if we would have to classify these animals as 'rational animals' (*animalia rationabilia*) by ascribing them the capacity of judging (as Wodeham claimed) we could still stress that humans 'are called rational in the proper sense of the term' (*proprie rationalis appellatur*).[18] For Gregory, this is still better than denying them the capacity of judging, because their sensory judgments are more than just an instinctual reaction that is automatically triggered when a particular stimulus is perceived. Like human judgments, the judgments of nonhuman animals can vary and this is how one can account for the fact that they sometimes react differently to the same stimulus: they simply judge differently about the same thing.

At first glance, this seems to make a pretty strong case against Wodeham's theory.[19] Nevertheless, Wodeham defends his explanation with the following argument. If we observe animals behaving differently towards the same thing,

---

idem, et non movetur ad ipsum. Ergo motus iste, cum sit per appetitum animalem et talis appetitus sequatur apprehensionem, praesupponit praeter simplicem apprehensionem sensibilis iudicium quo iudicatur illud utile vel necessarium aut tale vel tale."

16 Perler (2006), 96.

17 Gregory of Rimini, *Lectura in primum et secundum Sententiarum*, dist. 1, q. 3, eds. Trapp & Marcolino (1981), 305: "Nec valet ratio cuiusdam doctoris [i.e., Adam Wodeham; A.O.], qui dicit quod secundum hoc in talibus animalibus esset intellectus practicus et deberent talia animalia appellari rationabilia. Non enim haec sequuntur loquendo de huiusmodi ad sensum, in quo dicimus hominem esse rationalem et habere practicum intellectum, quia homo non tantum singularia iudicia sed et universalia, nec tantum circa sensibilia sed et circa insensibilia habet iudicia, et ex talibus ratiocinatur et proprie rationalis appellatur."

18 Ibid.

19 This is also noted by De Haan (2014), 423, although he does not discuss Wodeham's but Aquinas' theory.

he claims, this is because they have different memories of the same things. That is to say, while a certain thing is remembered as pleasant by one animal, another member of the same species might remember this thing as harmful because it has experienced it differently in the past. This then creates different responses to like stimuli.[20] For instance, some dogs run away from sticks, whereas others rejoice in retrieving them. But can this argument really refute Gregory's objection?

Indeed, it seems that this defence only applies to cases in which different individuals of the same species (e.g. dogs) react differently to the same stimulus (e.g. sticks). If, however, one and the same animal constantly changes its reaction, like in the example of eating bread, it is rather unlikely that a change in memory is responsible for this. Only a difference in judgment can account for this difference in behaviour, Gregory argues. Wodeham does not find this argument convincing, by contrast, because even humans do not always act on the basis of judgments, as he tries to show with the following example. If my head is itching, I may shave it because I remember shaving to be a pleasant remedy in this case. But this action is not the result of a 'complex apprehension' (*apprehensio complexa*). Rather, I do this without 'any deliberation or previous judgment' (*sine omni deliberatione vel iudicio praevio*).[21] This could also be applied to the animal that sees a piece of bread: if it is hungry, it goes for the bread, but if it is satiated it ignores it, just as I do not shave my head unless it itches.

Regardless of whether one finds Wodeham's arguments convincing or not, his debate with Gregory of Rimini nicely shows how very different explanations can be given for the same kind of animal behaviour (which is something

---

20 See Adam Wodeham, *Lectura secunda in primum librum Sententiarum*, prol., q. 4, §8, ed. Wood (1990), 100: "Ubi autem unum individuum speciei fugit aut prosequeretur praesens vel absens quod aliud individuum [eiusdem speciei nec] prosequeretur [nec fugit], hoc est ex memoria et imaginatione poenae vel delectationis alias ex consimili obiecto proveniente, ita quod ibi etiam nulla est talis compositio obiectiva. Sed ad positionem apprehensionis rei praesentis cum imaginatione vel memoria simplici poenae vel delectationis praehabitae, statim prosequitur vel fugit."

21 See ibid.: "Nec est hoc inconveniens bruto, cum etiam in nobis experiamur, cum ad solam positionem simplicium apprehensionum multa frequenter facimus sine omni alia deliberatione vel apprehensione complexa, sicut scalpimus caput cum sentimus pruriginem et fricamus barbam. Et alios multos gestus et motus habemus sine omni deliberatione vel iudicio praevio. Item, licet forte ad hoc quod virtuose diligamus aliquid, praesupponitur iudicium vel deliberatio, tamen ad hoc quod absolute diligamus, frequenter sufficit simplex apprehensio. Quare igitur non eodem modo ad hoc quod brutum appetatur vel appetat?"

we shall return to in Part 6). Moreover, it illustrates that some medieval authors had no problem with ascribing the capacity of judging to nonhuman animals, while others, like Wodeham, took this to involve a sort of deflation of the concept of judging and an anthropomorphic way of speaking. To be clear, none of the other authors covered in this part denied that there are significant differences between the way in which we judge and the way in which nonhuman animals judge. Most importantly, they agreed that the judgments of nonhuman animals do not have a predicative structure. But they were sympathetic to the idea that such sensory judgments are a form of judgments too and, as has been shown, they range from very basic natural judgments in Aquinas to more complex forms, such as Olivi's reflexive judgments. What remains to be seen is whether nonhuman animals can also combine various judgments and form syllogisms or arguments. In other words, the question is whether they can engage in the third intellectual operation: reasoning. This question will be the subject of the following part.

PART 4

Reasoning

Introduction to Part 4

Reasoning or syllogising is what comes third in the triad of intellectual operations. Once we have formed concepts and once we have arranged these concepts in a certain order, that is, in a judgment or in a proposition, we can take a number of judgments and arrange them in a certain order as well. The result is what is usually called a *syllogism*. And, briefly speaking, the process of arranging propositions in such a way that they form a syllogism can be called *reasoning*. From a modern perspective reasoning might seem like a broader concept than syllogising. And so one might say that reasoning is not necessarily syllogistic. But for the medieval authors we are concerned with, 'syllogising' (*syllogizari*) and 'reasoning' (*ratiocinari*) are largely synonymous and thus often used quite interchangeably. The crucial question, however, is whether they saved those terms for describing certain cognitive processes of human beings or whether they considered the possibility that nonhuman animals engage in reasoning too.

The present part sets out to explore their views on animal reasoning.[1] At first glance, this might seem to be an odd aim. Although, as we have seen in the previous part, many of the authors covered here ascribe certain types of judgments to nonhuman animals, it seems questionable whether those judgments could feature as parts of a syllogism, because they do not consist of general terms or concepts. In other words, they are not, strictly speaking, propositions. It seems doubtful, therefore, that nonhuman animals are capable of reasoning at all. Despite such doubts, the question of whether other animals reason too was taken to be a serious question by various medieval philosophers since they knew of several cases in which dogs, cats, apes, or other nonhuman animals at least appeared to reason. By and large, these cases could be summarised as examples of problem solving. The examples are puzzling insofar as problem solving seems to involve some cognitive flexibility and intelligence. If one observes an animal solving a problem, one is usually inclined to say that it does not do this by instinct, but by reasoning because, like us, it first seems to analyse the situation and then tries to figure out how to deal with it. We are often impressed by animals' capacity for problem solving because the way in which they do this appears to be very human-like. But what is even more fascinating than the behaviour itself is the question of what brings it about.

1 For a comprehensive overview of animal reasoning from Antiquity to the early modern era, see Roling (2011). See also Köhler (2014), 375–389.

 | DOI 10.1163/9789004363779_024

What kind of cognitive processes underlie it? Is it really reasoning that is involved or do nonhuman animals employ very different cognitive strategies for solving problems?

In the following chapters, we will see that there is a whole spectrum of answers to those questions among various thirteenth- and fourteenth-centuries authors. They range from the position that those animals that lack intellectual faculties are totally incapable of reasoning to the view that the lack of intellect and reason does not make reasoning entirely impossible. We shall begin with authors who are sympathetic to the former view such as Thomas Aquinas, Gregory of Rimini, and John Duns Scotus (Chapter 19). Roger Bacon who comes next also goes in this direction, but his arguments are somewhat different as he develops alternative explanations for seemingly rational behaviour in nonhuman animals (Chapter 20). Chapters 21 and 22 will finally reconstruct and discuss the views of Albertus Magnus, John Buridan, and Nicole Oresme who seem to leave more room for reasoning in nonhuman animals.

CHAPTER 19

Quasi-Reasoning (Thomas Aquinas, Gregory of Rimini, John Duns Scotus)

Thomas Aquinas addresses the question of whether or not nonhuman animals reason in connection with the famous example of Chrysippus' dog, which was originally formulated by the Stoic logician of the same name.[1] Like most other Stoics, Chrysippus was relatively skeptical about, if not even hostile to, the idea of animal rationality.[2] His actual intention was not to make a case for the rationality of dogs. Rather, he simply aimed to illustrate a certain type of syllogism, namely, the so-called disjunctive syllogism. And so he told the story of a hunting dog that chases an animal but then loses sight of it and comes to a fork with three paths. Since the dog cannot see where the animal went it sniffs around the first and then around the second path, before finally taking the third way without sniffing. The dog in the example thus seems to employ a syllogism of the form 'The animal took either A or B or C, so if it did not take A or B it must have taken C'. As has been stated, Chrysippus simply took this to be a good illustration of a disjunctive syllogism. However – and quite ironically, one could say – his example became one of the most widespread examples of animal reasoning, especially through the *Outlines of Pyrrhonism* in which Sextus Empiricus turned this example against the Stoics' denial of animal intelligence and made it a proof of their rationality.[3]

Thomas Aquinas, like many other medieval authors, likely knew this example through writings of the Church fathers, such as Basil of Caesarea's and Ambrose of Milan's *Hexaemeron*.[4] But what matters here is not so much how this example was handed on from one generation to the next but rather how

1 On the origin, history, and reception of this example, see Floridi (1997). A recent philosophical analysis and discussion provides Rescorla (2009).

2 On the Stoics' position see Dierauer (1977), 205–238; Sorabji (1993b), 20–28; Steiner (2005), 77–92.

3 See Sextus Empiricus, *Outlines of Pyrrhonism* I.14, §69, tr. Bury (1933), 41–43.

4 See Basil of Caesarea, *Homiliae IX in Hexaemeron* IX.4, ed. Migne (1886), 198B-C: "Et quidem canis expers rationis est attamen sensum rationi haud imparem habet. Quae enim mundi sapientes per multum vitae otium desidentes vix invenerunt, videlict ratiocinationum ambages, haec ab ipsa natura edoctus canis perhibetur." Ambrose of Milan, *Hexaemeron libri sex* VI.4, §23, ed. Migne (1882), 265D-266B. On the reception among Christian writers in particular see Moretti (2005) and Lauzi (2012), 217–219.

 DOI 10.1163/9789004363779_025

someone like Aquinas interpreted it. Aquinas mentions the example of the dog in the *Summa theologiae* when he asks whether nonhuman animals are capable of making a 'choice' (*electio*). In his opinion, the dog's behaviour is a good example for 'the astonishing sagacities in the works of animals' (*mirabiles sagacitates in operibus animalium*).[5] This view, of course, very much depends on the fact that the dog 'takes the third path without further exploration' (*per tertiam viam incedit non explorando*), as Aquinas puts it. He thus stresses the fact that the dog takes the third path without sniffing at it, as he did with the other two. Now, one could object that it is unlikely that the dog would not sniff at the third path, too, and thus that the example is flawed from the very beginning. Despite this objection, it is significant that the majority of thinkers accepted the case as a plausible description of canine behaviour. The key matter of debate, then, is whether or not such an animal takes the third path by syllogising.

According to Aquinas, the dog takes the third path 'as if using a disjunctive syllogism' (*quasi utens syllogismo divisivo*). This means that it looks as if the dog reasoned in this situation. But actually it employs neither a disjunctive syllogism nor any other kind of rational argument, Aquinas argues. The reason simply is that dogs lack intellect and reason, therefore they cannot engage in reasoning. But how else can one account for the dog's behaviour? Aquinas gives the following reply: like many other nonhuman animals, the dog has 'a natural inclination to certain most orderly processes' (*inclinatio naturalis ad quosdam ordinatissimos processus*).[6] Admittedly, this might seem like a relatively obscure description of what is going on in the dog's head while it is standing at the fork in the road, but Aquinas' idea is that the dog's behaviour is based on a comparatively complex cognitive process. However, this process is an orderly process only by virtue of 'the highest art' (*summa ars*), that is, the

5 Thomas Aquinas, *Summa theologiae* I–II, q. 13, a. 2, co., ed. Leonina VI (1891), 99: "Et hoc etiam sensui manifestum videtur, apparent enim mirabiles sagacitates in operibus animalium, ut apum et aranearum et canum. Canis enim insequens cervum, si ad trivium venerit, odoratu quidem explorat an cervus per primam vel secundam viam transiverit, quod si invenerit non transisse, iam securus per tertiam viam incedit non explorando, quasi utens syllogismo divisivo, quo concludi posset cervum per illam viam incedere, ex quo non incedit per alias duas, cum non sint plures. Ergo videtur quod electio brutis animalibus conveniat."

6 Ibid., 100: "Et ex hoc contingit quod in operibus brutorum animalium apparent quaedam sagacitates, inquantum habent inclinationem naturalem ad quosdam ordinatissimos processus, utpote a summa arte ordinatos. Et propter hoc etiam quaedam animalia dicuntur prudentia vel sagacia, non quod in eis sit aliqua ratio vel electio. Quod ex hoc apparet, quod omnia quae sunt unius naturae, similiter operantur."

divine creator. One could perhaps compare this to the processes of a complex machine, say, a computer. The computer calculates and solves all kinds of difficult tasks, but it does this only because this is how it has been constructed and programmed by a rational human being. Hence, the computer itself would not qualify as a rational being in the strict sense of (faculty and process) rationality. Similarly, the dog manages to find the right path, because this is how it has been programmed, so to speak, by the divine creator. Still, the natural inclination from which arises the 'orderly process' in which the dog engages makes this a very different process than the syllogistic reasoning that brings about the choice of a human being.[7] In this sense, the dog is closer to a machine than to a human.

One might find this comparison inappropriate for two reasons. First, it might be thought to be anachronistic. Second, it seems to turn Aquinas into a Cartesian who considers nonhuman animals to be soulless machines.[8] But neither is the case. As far as the charge of anachronism is concerned, it is important to note that Aquinas himself draws a comparison between cleverly constructed machines and nonhuman animals. His point is that many things, including, for instance, arrows or clocks, 'are moved by reason' (*moventur a ratione*). But this does not mean that they are rational beings themselves.[9] Rather, an arrow is brought into motion by a rational agent, and so if it hits its target this is because the human being that launched it is a good archer and not because the arrow itself is capable of moving into the right direction. It is only 'as if' (*ac si*) the arrow itself knew where to go to. Similarly, clocks display the time because this is how they have been constructed by some rational human craftsman. But what about animals? In contrast to arrows or clocks, they do have bodily organs and sensory powers by which they move. So in this respect they are definitely superior to non-cognitive beings, which is something Aquinas does not deny. Admittedly, Descartes did not deny that nonhuman animals have certain

7 This is exactly the argument Michel de Montaigne criticises in his *Essais* II.12, 168: "Je dy donc, pour revenir à mon propos, qu'il n'y a point d'apparence d'estimer que les bestes facent par inclination naturelle et forcée les mesmes choses que nous faisons par nostre choix et industrie." On this see also Wild (2006), 79.

8 This is what Barad (1995), 12, insinuates in her interpretation of this passage.

9 Thomas Aquinas, *Summa theologiae* I–II, q. 13, a. 2, ad 3, ed. Leonina VI (1891), 99f.: "[…] in omnibus quae moventur a ratione apparet ordo rationis moventis, licet ipsa quae a ratione moventur rationem non habeant: sic enim sagitta directe tendit ad signum ex motione sagittantis, ac si ipsa rationem haberet dirigentem. Et idem apparet in motibus horologiorum et omnium ingeniorum humanorum quae arte fiunt. Sicut autem comparantur artificialia ad artem humanam, ita comparantur omnia naturalia ad artem divinam." Descartes also compares animals to clocks, see Wild (2006), 161.

cognitive capacities either. But, in his view, the possession of these capacities does not justify the ascription of a soul to their possessor.[10]

Aquinas obviously disagrees on this point.[11] Nevertheless, he emphasises that 'just like artificial things compare to human art, so do all natural things compare to the divine art'.[12] In other words, every creature owes its design and functionality to the divine creator. But there are, of course, crucial differences in design and functionality. Most importantly, only human beings are endowed with intellect and reason, as well as with will. Because of this, only humans are capable of reasoning and only they can make a choice, according to Aquinas.[13] Thus, the dog at the fork does not reason freely. It does not arrange premises and conclusions and it does not come to a decision by ruling out two of three options, as we would. Rather, it owes the capacity to find the right path without sniffing to the divine creator, just as an arrow owes its capacity to hit the target to the archer or as a calculator owes its capacity to solve various mathematical problems to a clever engineer. To be clear, this does not necessarily make these beings inferior to rational human beings. As we all know, calculators usually calculate much more quickly than we do. Similarly, many animals might have fewer difficulties in solving certain problems even though they do not possess intellect and reason.[14] Nevertheless, how they do this is very different from how we do it.

In Aquinas' view, another sign of this difference is that all members of a nonhuman animal species act alike. There is a uniformity in their behaviour: all swallows build their nests in the same way and, similarly, all spiders construct their webs in a very similar manner.[15] Many, if not all, of the things 'they

---

10 For a thorough analysis of Descartes' position see Wild (2006), 135–210. See also Cottingham (1978).

11 On the differences between Descartes and Aquinas in this context see also Davids (2017), 192f.

12 See n9 above.

13 Floridi (1997), 43, obviously misreads Aquinas here as he states that Aquinas "held that the episode [of Chrysippus' dog; A.O.] proved that animals too have the capacity of choosing." On this misinterpretation see also Davids (2017), 191.

14 This point is not made by Aquinas but by Basil of Caesarea and Ambrose of Milan; see n4 above.

15 See Thomas Aquinas, *Summa contra gentiles*, lib. II, q. 82, co., ed. Leonina XIII (1918), 513: "In animabus autem brutorum non est invenire aliquam operationem superiorem operationibus sensitivae partis: non enim intelligunt neque ratiocinantur. Quod ex hoc apparet, quia omnia animalia eiusdem speciei similiter operantur, quasi a natura motae et non ex arte operantes: omnis enim hirundo similiter facit nidum, et omnis aranea similiter telam."

do as if they were moved by nature' (*quasi a natura motae*), as Aquinas states. But why does he say that they act '*as if* they were moved by nature'? Why does he not simply say that they are moved by nature? Aquinas himself does not explain why he has chosen this particular phrasing. But the most likely explanation is that he is aiming, once more, to stress the difference between cognitive and non-cognitive beings. Non-cognitive beings are entirely moved by nature. If a stone falls to earth, this happens not because the stone cognitively responds to a certain stimulus. Rather, it happens because of certain natural forces. Similarly, an arrow does not fly because some cognitive power has triggered its movement. Rather, it moves because a cognitive agent has set it into motion. Nonhuman animals, by contrast, do (most of) the things they do because they perceive something and then react to what they perceive. The dog does not take the right path because of the laws of gravity or because a human being pushes it in this direction. Instead, its behaviour is based on a cognitive process. Nevertheless, this process is different from the process that occurs in a human being in that it is not a *rational* process. Hence, even though it sometimes looks as if other animals reason, they can not be credited with that capacity, according to Aquinas.

Like Thomas Aquinas, Gregory of Rimini and John Duns Scotus also think that nonhuman animals do not actually engage in reasoning. For Gregory, the behaviour of dogs shows, on the one hand, that having 'complex cognition' (*notitia complexa*) is nothing exclusively human. In his view, these dogs 'judge' (*iudicant*) that their prey must have taken the remaining path. But, on the other hand, they only 'judge as if they reasoned' (*quasi arguentes iudicant*). So, like Aquinas, Gregory thinks that the cognitive process that makes the dog take the remaining path does not qualify as reasoning.[16] It might seem somewhat incoherent to claim that Gregory belongs to Aquinas' camp, especially because we have seen him attacking Adam Wodeham's attempt to account for animal behaviour on the basis of simple stimulus-response models before (see Chapter 18).[17] Nevertheless, his phrasing is quite clear when it comes to the question

16 See Gregory of Rimini, *Lectura super primum et secundum Sententiarum*, dist. 3, q. 1, eds. Marcolino & Trapp (1981), 305: "Tertio, idem probatur ex experientia quae de canibus venaticis fertur, videlicet quod, cum ferram fugientem insequuntur, si ad bivium veniant, odorant unam viarum, in qua, si non percipiant odorem fugientis, subito absque alterius viae odoratione per alteram pergunt. Quod non videtur aliunde contingere, nisi quia iudicant feram non transisse per illam quam odorati sunt, et ex hoc quasi arguentes iudicant transisse per aliam." The opposite view is held by William of Ockham, *Quaestiones in librum quartum Sententiarum (Reportatio)*, q. 14, eds. Wood & Gál (1984), 315.

17 Adam Wodeham also mentions the syllogising dog in his *Lectura secunda in primum librum Sententiarum*, prol., q. 3, §2, ed. Wood (1990), 67.

of reasoning. For even though he thinks that nonhuman animals have complex cognition, he holds that this does not amount to a syllogistic process. Furthermore, he shares Wodeham's doubt that we cannot naturally know what kind of cognition other animals have. All we can do is to observe their behaviour and to infer what kind of cognitive process is likely to underlie this behaviour.[18]

This epistemological caveat is also stressed by John Duns Scotus. As he points out, there are many cases in which the behaviour of humans and nonhuman animals looks very much the same. The example he gives is the following: if a human being has to make a choice between two different paths, she is likely to take the shorter one. This decision results from a syllogistic argument of the form 'By taking the shortest path I arrive much faster at my destination; this one is shorter, therefore, I take it'.[19] So, in short, our behaviour is based on what Scotus calls a 'discourse of reason' (*discursus rationis*). But what about a dog, for instance? If seen from the outside, the dog's behaviour is relatively similar to our own behaviour because if it has to make a choice between two paths, it usually takes the shorter one, too. However, the cognitive process involved in this situation is (or might be) very different. Whereas humans take some time to deliberate on what would be the better option to choose, the dog chooses the shorter path 'by sensitive appetite alone' (*ex solo appetitu sensitivo*), as Scotus puts it.[20] So even though there is no significant difference in behaviour, there is, in Scotus' view, a crucial difference with regard to what brings about this behaviour.

Two objections could be raised against Scotus' explanation. First, if we say that the dog chooses the shortest path by some very simple sensory mechanism why should we assume that the analogous human mechanism is much more sophisticated? Maybe humans also choose the shortest path by some

18 Gregory of Rimini, *Lectura super primum et secundum Sententiarum*, dist. 3, q. 1, eds. Marcolino & Trapp (1981), 304: "Et hic dico quod, quamvis istud non possit nobis plene naturaliter esse notum, probabiliter tamen ex his, quae percipimus, possumus arguere quod talem notitiam habeant."

19 See John Duns Scotus, *Quaestiones super libros Metaphysicorum Aristotelis*, lib. 1, q. 3, §10, eds. Andrews et al. (1997), 89: "[…] bruta multa faciunt ex cognitione eodem modo quo fierent ab homine cognoscente per discursum rationis; ergo videntur habere similem cognitionem. Antecedens patet: sicut homo syllogizans argueret 'per breviorem viam pervenitur ad intentum, haec est brevior, ergo etc'. et ex tali discursu breviorem eligeret ad aliquid obtinendum, similiter videtur canis eligere prosequendo leporem, et sic in aliis."

20 Ibid., §11: "Ad hoc dicitur quod licet similiter agant quaedam, sicut homo ex deliberatione ageret, non tamen oportet quod similem cognitionem habeant. Nam illud, quod eligitur ex deliberatione, posset etiam idem non eligi ex deliberatione, sed ex solo appetitu sensitivo."

sort of instinct? Second, if all we can do is to observe animal behaviour, how can we be so sure that we draw the right conclusions about the underlying cognitive processes? Maybe dogs and other nonhuman animals reason as well. To the first objection Scotus would possibly respond that the discourse of reason is something we can experience by introspection. If we monitor what is going on in our head when we choose a path, we are likely to see that we are reasoning in this situation. Moreover, other people might tell us that they usually reason in such a situation, and so we might conclude that we reason too because we belong to the same species of animal. Finally, Scotus might also refer to the teleological aspect of the possession of reason. Since we possess intellect and reason, he could say, it is unlikely that we do not use them. Otherwise it would be unclear why we have them at all, since 'nature does nothing in vain', as a popular Aristotelian saying has it.[21] There is no guarantee that we *always* use them, and Scotus does not doubt that many cognitive processes such as sensory perception do not require intellectual faculties. Nevertheless, we have them; therefore, we must use them often.

Nonhuman animals, by contrast, do not have such faculties, as Scotus would reply to the second objection. Consequently, they cannot engage in reasoning. All they have is a certain number of external and internal senses, and so if they do something, it is not rational discourse that is behind this behaviour but sensation. Now, one could say that this is not a very convincing argument because it rests on the metaphysical assumption that nonhuman animals lack certain faculties of the soul. But why should we agree to this premise? Unless we accept that God created both human and nonhuman animals and gave intellect and reason only to the former, Scotus' argument does not work. On the one hand, that is a very strong objection indeed. But on the other hand, Scotus could defend his view without referring to the metaphysical premise, namely, by referring to Morgan's Canon. As we have seen in Chapters 12 and 18, this modern methodological principle recommends favouring explanations that refer to low-level cognitive processes over explanations that refer to higher-level processes as long as the former can coherently account for an animal's behaviour.[22] With this principle, Scotus could easily defend his argument that dogs find the shortest path by sensitive appetite while we find it by reasoning.

---

21 See *Les Auctoriates Aristotelis*, ed. Hamesse (1974), n. 168, 188: "Natura nihil facit frustra, unde non deficit in necessariis, nec abundat in superfluis." This sentence was commonly evoked by medieval Aristotelians. Note, however, that Scotus does not mention it in the present context.

22 See p. 92 n10.

Still, the question is whether this explanation also applies to other cases of animal behaviour. Maybe there are cases in which sensory appetite is too simple a mechanism in order to account for what an animal does. So the question is, how can one coherently explain more complex behaviours, especially, if one aims at saving the metaphysical premise according to which only humans possess intellect and reason? Not only Scotus, but also Aquinas and Gregory of Rimini might have answers to these questions. But since they do not address them explicitly, answering them on their behalf would be rather speculative. Therefore, it is much more promising to look at another thinker of the same period, namely, Roger Bacon who is explicitly concerned with complex animal behaviours and how one can account for them without giving up the abovementioned metaphysical premise.

CHAPTER 20

Quasi-Reasoning and Cogitation (Roger Bacon)

In comparison to other medieval accounts of animal reasoning, Bacon's account is quite comprehensive in addition to being original. As with the case of universals and concepts, the context of his discussion is Alhacen's theory of different modes of vision (see Chapter 11). The main question is whether nonhuman animals have only what is called 'cognition by sense alone' (*cognitio solo sensu*) or whether they also possess 'cognition by knowledge' (*cognitio per scientiam*) and 'cognition by syllogism' (*cognitio per sillogismum*). The latter kinds of cognition in particular give rise to some questions, Bacon says, because if one applies the terms 'knowledge' and 'syllogism' in the way in which they are usually applied in logic, natural philosophy, and metaphysics, one cannot employ them for the cognition of nonhuman animals. The reason is that knowing and syllogising are usually taken to be intellectual operations. Hence, those animals that lack intellect and reason cannot engage in knowing and syllogising.[1]

Quite obviously, this argument rests on the assumption that the possession of intellectual powers is what distinguishes us from other animals. But how can we know that this definition of the animal/human boundary is correct? As said before, we cannot look into their heads. Therefore, the only way to test this definition is to compare the behaviour of humans and nonhuman animals. If it turns out that we can coherently account for the behaviour of nonhuman animals without referring to intellectual powers, there is no reason to revise the common metaphysical premise substantially. But if there are cases in which such an explanation reaches its limits, a revision of this premise might be required. In particular, such cases involve animals that manage to solve certain problems because problem solving seems to require the capacity to reason. There are undoubtedly many situations in which animals solve problems by instinctual actions or because of some peculiar physical feature: in the broad sense of the term, swallows, for example, solve the problem of finding shelter by building a nest. But if we see nonhuman animals dealing with novel situations, the question is whether this can be explained by instinct alone.

1 See Roger Bacon, *Perspectiva*, pars II, dist. 3, c. 9, ed. Lindberg (1996), 246: "Et si accipiamus 'scientiam' et 'sillogismum' sicut in logicalibus et naturalibus et metaphysicis, ut est in usu vulgi philosophantium, necesse est quod sit anima rationalis, quia sillogismus et scientia pertinent ad eam solam, ut accipiuntur in dictis scientiis."

 | DOI 10.1163/9789004363779_026

Bacon presents a whole range of examples of this sort, but the most illuminating one is the following. He claims to have once observed a cat that tried to catch fish which were swimming in a large stone container. However, the cat had a hard time grabbing ahold of one of them, because they always escaped its claws in the water. Nevertheless, it found a way to solve this problem: it simply pulled the stopper out of the container, waited for the water to drain out, and then took the fish from the dry ground.[2] This is a very clear illustration of problem solving in a nonhuman animal. The cat has a certain goal, encounters certain obstacles while trying to reach this goal, but finally manages to overcome these obstacles. Still, the most decisive question is: by virtue of what cognitive process does the cat actually do this? More precisely, one needs to ask whether this cognitive process is the same as the one by which we would solve the very same or a similar problem.

The simplest explanation for the cat's behaviour would be that it results from pure chance: the cat accidentally pulls out the stopper, and so in fact neither analyses the situation nor understands anything about the (causal) relations between claws, fish, water, stone containers, and stoppers. Rather, it continues to try to catch the fish and then pulls the stopper out by chance without actually intending to. In modern research, one would try to falsify such an explanation through empirical testing. This would involve, for instance, confronting the cat with this problem again and again to see whether it always pulls out the stopper. In addition, one would confront it with other problems, as well, especially those that require more planning and strategic acting and thus cannot be solved by pure chance. If it manages to solve these problems too, the explanation that it pulled the stopper out of the stone container by accident would become more and more unlikely. Instead, one would then be inclined to say that the cat found the solution to this problem by some form of reasoning, which one would need to further specify.

While Bacon does not consider the explanation that the cat accidentally pulls out the stopper, he discusses at some length another possible explanation, namely, the theory that the cat's behaviour is based on a process of reasoning. This explanation is quite appealing as the cat obviously has a certain aim – it aims to catch the fish – and 'because of this aim which resembles a conclusion it gathers many things in its cogitation which resemble premises',

2 Ibid., 248: "Atque vidi murilegum qui desideravit pisces natantes in magno vase lapideo, et cum non potuit propter aquam deprehendere eos, abstraxit clepsedram et deduxit aquam donec vas siccabatur, ut in sicco pisces reperiret; plura igitur opera hic concepit ut finem intentum haberet."

as Bacon puts it.[3] In fact, we can easily describe the cognitive process underlying the cat's behaviour in terms of a syllogistic argument. This argument has a number of premises, such as, 'If there is no water in the container, one can grab the fish from the dry ground' and 'If one pulls the stopper out the water drains out'. This leads to the conclusion 'Therefore one must pull out the stopper', and that is exactly what the cat does. This argument is not a theoretical, but what is usually called a practical syllogism, that is, a syllogism in which the conclusion is an action. Nevertheless, the conclusion is the result of an inferential process, and so the cat obviously has 'cognition by syllogism'.

There is, however, a serious problem with this explanation, as Bacon points out. In his view, it is wrong to call this kind of cognition 'cognition by syllogism' because 'the disposition of an argument in a certain form and the distinction between the conclusion and the premises pertain to nothing but the rational soul'.[4] And since nonhuman animals do not have rational souls, they are incapable of forming and arranging syllogisms. It is only, he says, 'as if they were by themselves inferring a conclusion from premises' (*ac si arguerent apud se conclusionem ex premissis*).[5] That is, he shares the view of Thomas Aquinas, Gregory of Rimini, and John Duns Scotus, who all hold that nonhuman animals cannot actually reason (see Chapter 19). And like them, he defends the metaphysical premise according to which nonhuman animals lack intellectual faculties: 'they use sense alone because they do not have an intellect', as he states in another passage of the *Perspectiva*.[6] Therefore, they only seem to reason or they 'quasi-reason', as others call it. So, like Chrysippus' dog, Bacon's cat does not actually run through a syllogistic argument; hence its action of pulling out the stopper is not actually the result of an inferential process.

But this explanation seems to beg the question. It takes for granted that the cat cannot reason simply because it lacks a rational soul. But was not the cat's success in problem solving what gave rise to doubts about this metaphysical premise? Consequently, it seems unsatisfying to argue that the cat does not reason because it lacks reason. However, one can maintain this premise as long

3 Ibid., 250: "[...] propter igitur hunc finem qui assimulatur conclusioni, multa colligit in sua cogitatione que premissis simulantur."

4 Ibid.: "Propter quod equivocatur hic, vel magis est vitium translationis, que non habet vocabulum proprium ad hunc modum cognoscendi. [...] Sed de argumento, oportet considerare quod dispositio argumenti in figura et distinctio conclusionis premissis non pertinent nisi ad animam rationalem."

5 Ibid: "Et sic est de infinitis in quibus bruta animalia cogitant multa per ordinem respectu unius rei quam intendunt, ac si arguerent apud se conclusionem ex premissis."

6 See *Perspectiva*, pars I, dist. 1, c. 4, ed. Lindberg (1996), 12: "[...] bruta animalia utuntur solo sensu, quia non habent intellectum."

as one can give a coherent alternative explanation for the cat's behaviour. That is exactly what Bacon tries to do. But unlike Scotus, he does not refer to a 'sensitive appetite'. Nor does he refer to 'natural inclination', like Thomas Aquinas. Instead, he argues that even though cats and other nonhuman animals cannot properly reason, they can engage in what he calls 'a certain collection of many things into one by natural industry and by instinct of nature' (*quedam collectio plurium ad unum ex naturali industria et instinctu nature*).[7] This might seem like a half-hearted compromise between an instinctual process, on the one hand, and reasoning, on the other hand. This is the case not only because Bacon emphasises the role of natural industry and instinct here, but also because he adds that the many things 'resemble premises' and the one thing 'resembles a conclusion'. Therefore, one could rightly argue that his theory is just another version of the theories of Thomas Aquinas, Gregory of Rimini, or John Duns Scotus. However, there is a crucial difference between the latter theories and Bacon's explanation.

The main difference is that Bacon ascribes the 'power of cogitation' (*virtus cogitativa*) to nonhuman animals. As mentioned in Part 2, Aquinas and most other medieval thinkers usually denied that this power can be found in other animals. They had no doubt that cogitation is a corporeal power located in the brain, but they took this to be an exclusively human faculty. Bacon, by contrast, makes cogitation the central point of his theory of animal cognition.[8] One might be inclined to think that this is a mere technicality, but it is actually more than that. The cogitative power is not just any inner sense, but a very peculiar faculty by virtue of which those beings who possess it can engage in what Bacon calls a 'discourse' (*decursus*; *discursus*). Such a discourse takes place when animals aim at a particular thing and then 'cogitate in an orderly way' (*cogitant per ordinem*) about how to get there.[9] The fishing cat is just one example among many, according to Bacon, since many animals "work on a means to ends basis," as Jeremiah Hackett put it.[10] Still, the question is, how shall we understand the difference between cogitation and instinct, on the one hand, and reasoning, on the other?

---

7 Roger Bacon, *Perspectiva*, pars II, dist. 3, c. 9, ed. Lindberg (1996), 248: "Sed quedam collectio plurium ad unum ex naturali industria et instinctu nature (que plura assimulantur premissis, et quod unum sit simile conclusioni, quia colligitur ex eis) bene potest reperiri apud bruta."

8 On Bacon's account of cogitation and animal cognition see also Sobol (1993), 114–117; Wood (2007), 39–45; Roling (2011), 244–246; Hackett (2013), 234–237; Köhler (2014), 371–373; Toivanen (forthcoming).

9 See n5 above.

10 Hackett (2013), 239.

As regards instinct, the most obvious difference from reasoning is found in the fact that the cat's response to the stone container is different from a sheep's response to a wolf. While the sheep instinctually flees wolves, it seems absurd to say that the cat has an instinct to pull out stoppers. Rather, pulling out the stopper is one step among others in order to achieve a certain end and the cat can make this step by virtue of the cogitative power. However, cogitation differs from reasoning in several respects, according to Bacon. First, the cogitative discourse is not arranged in particular modes and figures, like syllogistic arguments. It does not follow the rules of traditional logic in which every syllogism belongs to a certain type of argument. Second, the cat (and any other nonhuman animal) cannot deliberately distinguish between the beginning and the end of its cogitation. In other words, it cannot distinguish between premises and conclusions. Third, it also lacks what, in the modern discussion, is usually called 'meta-cognition', that is, it has no awareness of the fact that a mental discourse is going on in its head.[11] In short, cogitation lacks many classical features of reasoning.

There are several reasons why one might not be entirely content, if not totally unhappy, with these distinctions. Most importantly, one might object that it remains somehow unclear how cogitation is going to work if it lacks the above-mentioned features, especially, the second. For how can the cat arrive at the desired end if it is incapable of distinguishing means from ends? At first glance, this seems to be a serious problem indeed. But it derives from an overestimation of what cogitation amounts to. What Bacon has in mind here is a relatively simple inferential process that takes place at the sensory level. Unsurprisingly, he adopts this idea from Alhacen. In *De aspectibus*, Alhacen argues that by the so-called '*virtus distinctiva*' (which Bacon equates with the cogitative power) we can engage in some very basic inferential cognition.[12] He gives the example of a child that is offered a green apple and a red apple.

---

11 Roger Bacon, *Perspectiva*, pars II, dist. 3, c. 9, ed. Lindberg (1996), 250: "Sed decursum sue cogitationis non disponunt in modo et figura, nec ex deliberatione distingunt ultima a primis. Nec percipiunt se huiusmodi discursum facere, quia ex solo instinctu naturali sic decurrit cogitatio eorum."

12 This is also noted by Dewender (2003), 155. However, like Smith (2010), 342, he wrongly claims that this power is identical with the so-called *ultimum sentiens* in the forepart of the brain. As Köhler (2014), 378, points out, in the thirteenth century Adam of Whitby also adopted the idea of a *virtus distinctiva*. According to Adam, nonhuman animals do not have reason but 'a power similar to reason by which they discern, cogitate, act intelligently as well as quasi-reason and deliberate'.

Without further ado, Alhacen says, the child chooses the red apple because this apple looks much nicer and riper than the green one, and hence promises to be tastier.[13] Now the cognitive process that makes the child choose the red apple differs from a full-fledged argument, according to Alhacen. First, it differs because the child 'does not reason by composing and ordering premises' (*non arguit per compositionem et ordinationem propositionum*). Second, the child reasons 'without any need for words' (*sine indigentia in verbis*).[14] And, third, its 'argumentation' (*argumentatio*) takes place 'in a very short amount of time' (*in tempore valde parvo*) which is why it usually goes unnoticed.[15] The child, Alhacen claims, would not even understand *what* reasoning is if someone explained to it *that* it reasoned. Nevertheless, its choice is the result of 'sensory reasoning', as one might call it.[16]

For Bacon, it is clear that this kind of cognitive process cannot only be found in pre-rational human beings but also in nonhuman animals. In his view, the example of the cat, as well as other examples of animal behaviour, provide sufficient evidence for that. Admittedly, his argument is not entirely waterproof because one could still say that, both in the case of the cat and in the case of the child, it is *we* who describe their cognition as some sort of inferential process. But, in fact, they do not reason at all. To this objection, Bacon could respond as follows: it is absolutely right to stress that what is going on in the cat or the child is not strictly speaking an inferential process, let alone reasoning, since reasoning is a much more complicated operation which requires intellectual powers. Nonetheless, mere instinct is too simple an explanation for such behaviour. Therefore, we need to find an operation that takes place at a level in between instinct and reason. In his view, the operations of the cogitative power meet this criterion because, on the one hand, they are much closer to the operations of reason than the operations of other inner senses. On the other hand, cogitation is a corporeal power that is distinct from the faculties of intellect and reason. Therefore, one can ascribe it to other animals without violating the common definition of the animal/human boundary.

This compatibility of cogitation with the common metaphysical premise according to which nonhuman animals lack intellect and reason is undoubtedly

13 See Alhacen, *De aspectibus*, lib. II, c. 3, §38f., ed. Smith (2001), vol. 1, 108. Bacon reproduces this example almost literally in *Perspectiva*, pars 1, dist. 10, c. 3, ed. Lindberg (1996), 156, and in *Questiones altere supra libros prime philosophie Aristotelis* I, ed. Steele (1932), 8.

14 Alhacen, *De aspectibus*, lib. II, c. 3, §28, ed. Smith (2001), vol. 1, 104.

15 See ibid., §26, 103.

16 On this idea see Sabra (1978). On the analogy between sensory and intellectual reasoning see also briefly Tachau (1993), 668.

one of the strengths of Bacon's theory because it allows him to account for relatively complex behaviours without substantially revising this premise. At the same time, he also avoids a deflation of the concept of reasoning since this is reserved for the operations of the rational soul. Still, for his theory to be more convincing, he would need to spell out more clearly how the cogitative discourse works. Perhaps one could assist him by referring to modern theories of non-linguistic thinking. In short, such theories claim that we can ascribe *thoughts* to pre-linguistic children as well as to non-linguistic animals.[17] An example of such non-linguistic thought would be the sort of calculation we make when thinking about whether some object, say our wardrobe, fits between two other objects, such as our bookcase and the bed.[18] In this situation we do not really calculate. We do not take measures of the wardrobe and the space between the bookcase and the bed and then reason that, since the space measures only 55 inches and our wardrobe measures 65 inches, it will not fit. Rather, we simply try to visually imagine whether or not it will fit (although we often realise in the end that it would have been much smarter to take measurements). It seems that many nonhuman animals can also engage in this kind of very simple *imagistic thinking*, although they do not have the slightest idea of the rules of logic.

Within the framework of medieval theories of cognition one might be inclined to say that thinking is an intellectual operation. However, the cogitative power can be interpreted as a "faculty of thought" too as, for instance, David Summers has suggested.[19] Undoubtedly, cogitative thought is different from intellectual thought, particularly for those animals in whom there is no connection between the cogitative faculty and the intellectual powers, as there is in human beings. Nonetheless, the cogitative power might, for instance, enable the cat to mentally represent and rearrange images of dry stone containers and fish lying on the ground without presently seeing such things.[20] In order to make this situation real, the cat need not know anything about premises, conclusions, modes, figures, rules of inference, and so forth. It must, of course, somehow understand that pulling out the stopper is a means by which one can

---

17 The most famous and influential theories are those by Bermúdez (2003) and Tomasello (2014). Against such theories see Barth (2010).

18 See Bermúdez (2003), 36.

19 Summers (1987), 198.

20 Here one could see a parallel to what Arabic logicians called 'poetic syllogisms'. These syllogisms contained at least one premise or conclusion with figurative statements and were intended to convince less capable people in the realm of religion; see Schoeler (2005) and (2013). However, there is no clear evidence that Bacon knew this idea and applied it to nonhuman animals.

empty the container. But this 'understanding' is as unconscious as the child's 'understanding' of the ripeness and taste of apples.[21] This is also why Bacon emphasises that animals' cogitation works instinctually to some extent. But since it consists in a 'collection of many things into one', as he puts it, it is superior to a basic instinctual reaction. Therefore, it provides an explanation of how nonhuman animals can engage in problem solving or other complex behaviours even though they do not possess intellect and reason.

21 One could also put it as does MacIntyre (1999), 24f., who says that animals act for reasons even though "they do not possess the linguistic resources for articulating and uttering those reasons."

CHAPTER 21

Imperfect Argumentations and Practical Syllogisms (Albertus Magnus)

Thomas Aquinas, Gregory of Rimini, John Duns Scotus, and Roger Bacon all tried to account for comparatively complex behaviours in nonhuman animals. As we have seen, they gave different explanations for what brings about such behaviours. According to Aquinas, it is a natural inclination. According to Gregory of Rimini, it is a kind of complex cognition. For Scotus, sensory appetite is the source and, for Roger Bacon, it is cogitation by virtue of which such animals solve problems. Despite these differences, their explanations all converge on one point: they all agree that the cognitive processes underlying those behaviours cannot be called reasoning. In their view, reasoning is a prerogative of intellectual powers. And since they were – for whatever reason – rather reluctant to ascribe such faculties to nonhuman animals, they also denied them the capacity of reasoning. But as this and the following chapter will show, there were also authors who did not refrain from literally ascribing this capacity to other animals.

The first of these is Albertus Magnus. To be clear, Albert also subscribes to the metaphysical premise according to which the possession of intellect and reason is what marks the dividing line between humans and other animals. Despite this clear division, he considers the possibility that at least some highly developed nonhuman animals, such as monkeys and pygmies, come very close humans, not only with regard to physiology but also with regard to cognitive capacities. As mentioned in previous chapters, Albert supposes that the genus of monkeys or apes (*genus symiarum*) in general and the species of pygmies in particular are characterised by a relatively strong power of estimation.[1] They, for instance, care for other members of their species (see Chapter 16), and pygmies even have some kind of language, according to Albert (see Chapter 9).

1 See Albertus Magnus, *De animalibus*, lib. XXI, tr. 1, c. 3, ed. Stadler (1920), 1331: "[…] et in hoc cognoscitur symia melioris esse aestimationis quam aliud animal." See also ibid., 1332: "[…] et ideo pro certo aestimativam haec animalia habent lucidiorem omnibus aliis animalibus: propter quod melius aestimant de fantasmatibus quam alia animalia et melius eliciunt intentiones acceptas cum fantasmatibus quam aliqua aliorum animalium, ita quod videntur aliquid habere rationi simile." On Albert's account of primates see, for instance, Janson (1952), 84–89, and Roling (2010), 494–496.

 | DOI 10.1163/9789004363779_027

The strength of their estimative power is also crucial for the question of reasoning. As far as monkeys are concerned, Albert claims that they do not reason in the sense of forming 'practical syllogisms' (*sillogismi operis*). However, they engage in what he calls 'imperfect argumentation' (*argumentatio imperfecta*).[2] But what does that mean?

Usually, 'in the theoretical disciplines' (*in contemplativis*), imperfect argumentations are incomplete syllogisms such as enthymemes and examples.[3] What characterises those types of syllogisms is that they lack a certain number of premises. While a complete syllogism consists of at least two premises, enthymemes and examples have only one premise. In an enthymeme, one immediately jumps from a premise such as 'All humans are rational animals' to a conclusion like 'Socrates is a rational animal'. What is missing here is the premise 'Socrates is a human being'. Once we add this premise, we have a complete syllogism. So we can easily restore incomplete arguments by adding the necessary premises. But, if we do not do this, we run the risk of drawing erroneous conclusions. Take, for instance, the conclusion 'Brunellus is a rational animal'. This we inferred from the premise 'All humans are rational animals'. Yet, how can we be sure that Brunellus is a human being? In fact, Brunellus is a donkey. Therefore, the premise 'Brunellus is a human being' is false. But this we fail to see as long as we ignore this premise. Now does Albert want to say that monkeys often draw such erroneous conclusions?

In short, the answer is no. The point Albert wants to make is that "[t]he imperfect argumentations of primates are analogous to *exempla* and enthymemes," as Jörg A. Tellkamp puts it.[4] But why are nonhuman primates' imperfect argumentations not *actual* enthymemes or examples but only *analogous* to such incomplete forms of syllogising? The first and most important reason is that even incomplete syllogisms involve universals.[5] For instance, the

2 Ibid., 1331: "[...] et est in talibus animalibus non sillogismus operis, sed argumentatio imperfecta."

3 Ibid.: "Et sicut in contemplativis entimema et exemplum sunt argumentationes imperfectae, quarum tamen imperfectio perficitur per reductionem ad sillogismum: ita in istis animalibus sunt imperfecti sillogismi operum qui non habent nisi fantasticam aestimationem de operabili vel appetibili et appetitum facientem impetum ad opus."

4 Tellkamp (2016), 121. See also briefly id. (2013), 218–220.

5 See Albertus Magnus, *De animalibus*, lib. XXI, tr. 1, c. 3, ed. Stadler (1920), 1331f.: "Sed in hoc differentia est quod entimema procedit ex ea quae est sub universali prout est in pluribus vel in omnibus. Fantastica autem aestimatio non est nisi ex eo quod videtur tantum: et ideo frequenter decipitur sicut ex sophismate accidentis: quia non procedit nisi ex hiis quae hic et nunc videntur, reputans similiter fugienda vel persequenda quibus idem accidit accidens secundum hic et nunc."

premise 'All humans are rational animals' involves the general terms 'human' and 'rational animal'. However, even comparatively highly developed nonhuman animals fail to form concepts, Albert says (see Chapter 9). All they have, in his view, is 'a fantastic estimation' (*fantastica aestimatio*). By means of this faculty, they can only grasp particulars, that is, things which are 'here and now' (*hic et nunc*). Of course, they not only perceive these particulars as being of a certain colour or as having a certain shape, but also as being dangerous or harmless, desirable or undesirable. So if a monkey sees a banana, it does not simply perceive a yellow curved thing, but it sees something edible. Still, it perceives only *this* or *that* banana, and hence cannot form a premise like 'All bananas are yellow'. The second reason why such imperfect argumentations are only analogous to enthymemes or examples is that their estimative perceptions concern only practical matters. This means that most nonhuman primates only attend to things if they have some more or less immediate practical relevance for them. Finding a banana or fleeing a predator is something that matters. Theorising about mortality or rationality does not matter.

Despite these differences, Albert says that such animals have 'imperfect practical syllogisms' (*imperfecti sillogismi operum*).[6] Unfortunately, he does not give an example of such a syllogism. But a situation in which it plays a role could look as follows. A monkey sees a yellow curved thing. By tasting it, the monkey finds out that this yellow curved thing tastes sweet. It then develops a desire for sweet yellow curved things. The next time it sees a yellow curved thing it goes for it. Admittedly, this is a relatively simple situation which does not seem to involve any kind of complex behaviour. Nevertheless, one could say that it involves an imperfect practical syllogism insofar as the monkey infers from the sweetness of one particular yellow curved thing to the sweetness of another yellow curved thing and so transfers its desire for the first yellow curved thing to the second.

This explanation of the cognitive process meets the criteria mentioned above. First, there are no universals involved in this process, because the monkey infers from the sweetness of *this* yellow curved thing to the sweetness of *that* yellow curved thing.[7] Second, its inference, if one may say so, is entirely practical. It does not theorise about the notions of curvedness or yellowness

6 See n3 above.

7 Tomasello (2014), 13, who gives a very similar example with bonobos and figs remarks that the bonobo's inference that what it sees is sweet is "based on a categorization of the encountered fig as 'another one of those' and the natural inference that this one will have the same properties as others in the category." As seen in Chapter 9, Albert does not clearly ascribe the capacity of categorisation to nonhuman animals. Still, he might agree with Tomasello that

but eats a yellow curved thing if it sees one. This kind of inferential process is, however, relatively prone to error, as Albert stresses. The reason is that it is entirely based on sensory information.[8] If a monkey sees a yellow curved thing, it immediately concludes that it is sweet. But it might err because it cannot be sure that *this* yellow curved thing is the same kind of thing as *that* yellow curved thing it has tasted before. Thus, it falls prey to the same kind of error we make when mistaking bile for honey. In our case, this error also derives from a malfunction of our inner senses. But in contrast to other animals, we can correct such errors via our intellect (see Chapter 16). Monkeys cannot do this, which is a further reason why their argumentations are imperfect.

There is, however, one species of nonhuman primates that is superior to others, namely, pygmies. As mentioned in Chapters 9 and 16, pygmies are superior not only because they possess a better faculty of estimation but also because they have what Albert calls 'a shadow of reason' (*umbra rationis*). This shadow of reason is a separate power which excels estimation: it does not simply elicit intentions but generates some kind of experience, according to Albert.[9] Since the pygmy 'shares in experience to a small degree' (*experimenti parum participat*), Albert maintains that it can form practical syllogisms in the sense that it reasons inductively. Still, the conclusions it draws by induction are not perfect because they do not reach the level of universals.[10] That is to say, even the reasoning of pygmies does not involve conclusions of the form '*All* X are F' which

---

a primate needs to perceive certain similarities between two things in this situation. Otherwise, the inference would not work.

8 See Albertus Magnus, *De animalibus*, lib. XXI, tr. 1, c. 3, ed. Stadler (1920), 1331: "Sed in omnibus hiis non movetur nisi ex fantasmate: et ideo frequenter errat sicut et alia animalia quia sicut diximus in antehabitis, ubi fantasticum intellectui non coniungitur, frequens incidit error [...]." See also n5 above.

9 See Albertus Magnus, *De animalibus*, lib. XXI, tr. 1, c. 2, ed. Stadler (1920), 1329: "Virtus autem illa animae quam umbram rationis quamdam vocavimus superius, innominata quidem est a philosophis, sed circumloquendo cognoscimus quod haec vis aliquid potentiae addit super aestimativam. Cum enim aestimativa brutorum iudicet de intentionibus quae cum sensibilibus accipiuntur, ista plus facit quoniam res fert huiusmodi intentionis ad memoriam et elicit expertum et utitur illo postea ad quae confert." On the concept of a shadow of reason see Resnick & Kitchell (1996), 59f.; Köhler (2008), 434f., and (2014), 199; Toivanen (forthcoming).

10 Albertus Magnus, *De animalibus*, lib. XXI, tr. 1, c. 3, ed. Stadler (1920), 1332: "Inductionis autem aliquid videtur participare pigmeus in hoc quod experimenti parum participat secundum sillogismum operativum quem practicum Graeci vocant, sed perfecte non inducit, quando non progreditur usque ad universalis acceptionem." This peculiarity is also stressed by Köhler (2014), 388, but he does not discuss how it is to be understood.

is also why they are altogether incapable of using examples (*exempla*).[11] But if they in fact fail to do this, one could rightly ask whether it makes sense to say that pygmies use inductions because induction usually involves generalising. We recognise, for instance, that *this* human being is mortal and from this we infer that *all* human beings are mortal. So if pygmies fail to draw such general conclusions, it seems questionable whether they actually use inductive inferences or practical syllogisms at all.

One way to address this question would be to say that, by participating in experience, pygmies come much closer to the level of universals than other species of nonhuman animals. Certainly, they neither form concepts nor grasp universals. Nevertheless, by virtue of their shadow of reason they gather some sort of general information about different kinds of things. Moreover, they might make fewer mistakes than monkeys, for instance. While monkeys simply go for any yellow curved thing, pygmies might come closer to us as they are able to realise that not all yellow curved things are necessarily sweet. The monkey will also notice that a particular yellow curved thing is sweet, while another one is not. But the pygmy is more capable of learning from this experience. Still, the question remains whether the claim that pygmies reason in the sense that they use practical syllogisms is justified. Could it be instead that Albert simply employs a much broader notion of reasoning than those thinkers we have been looking at before? And is it on the basis of this broader notion that he can easily ascribe inferential capacities to pygmies, if not even to monkeys, too?

Two points are important to note here. First, it seems indeed as if Albert's ascription of reasoning to pygmies is based on a comparatively broad concept of reasoning. Even though pygmies have the same cognitive capacities as mentally disabled humans (*moriones*), according to Albert, they differ in one important respect: while mentally disabled human beings suffer from an abnormal 'lack of the use of reason' (*privatio usus rationis*), pygmies have 'a lack of reason by nature' (*privatio rationis ex natura*).[12] This means that their souls have a different nature than the souls of human beings and so they will

11 See *De animalibus*, lib. XXI, tr. 1, c. 3, ed. Stadler (1920), 1332: "Exemplo autem haec animalia nullo modo utuntur: eo quod exemplum fieri non potest sine aliqua rationis collatione." See also Roling (2011), 233.

12 Albertus Magnus, *De animalibus*, lib. XXI, tr. 1, c. 2, ed. Stadler (1920), 1328: "Et ideo pigmeus nichil omnino percipit de rerum quiditatibus nec umquam percipit habitudines argumentorum: et sua locutio et [corr. *est*; A.O.] sicut locutio morionum qui naturaliter stulti sunt eo quod non perceptibiles sunt rationum. Sed in hoc est differentia quod pigmeus habet privationem rationis ex natura, morio autem habet per accidens ex melancolia vel alio accidente non privationem rationis, sed potius privationem usus rationis."

never attain the level of (normal) human cognition. Mentally disabled humans at least have the potential for this. The pygmy 'neither cognises the quiddities of things nor the form of argumentations', Albert says, and this is something unchangeable, it seems, because one cannot 'heal' the pygmy's lack of reason. Therefore, pygmies are naturally closer to nonhuman than to human animals.[13]

Second, Albert is keen to stress that no other nonhuman animal species comes as close to our human species as this one. Nonhuman primates in general are 'likenesses of humans' (*similitudines hominis*), he maintains, not only because their physiology resembles ours, but also because their cognition is very human-like.[14] And pygmies in particular are 'almost midway' (*quasi medius*) between humans and other animals, especially because of their experimental cognition.[15] As seen above, this participation in experience is also the reason why they can engage in practical syllogising, in Albert's view. Thus, he would presumably agree with Roger Bacon that cats, for instance, cannot reason. But pygmies rank much higher on the ladder of beings than cats since they possess 'something similar to reason' (*aliquid rationi simile*), as Albert puts it.[16] Surely, there is a crucial difference between reason and the faculty they have: intellect and reason are immaterial faculties, whereas the shadow of reason belongs to the sensory, that is, the material soul. Nonetheless, the latter can bring about certain basic forms of reasoning. Those forms do not, of course, involve general terms and they remain entirely related to practical matters. Still, they have a syllogistic structure and lead to certain actions. In this sense, the pygmies' reasoning is not quasi-reasoning but reasoning in the true (though basic) sense of the term.

13 Ibid., 1329: "[...] magis tamen secundum naturam vicinus est bruto quam homini ut patet per antedicta: quoniam expertum magis est ad universale et contemplationem, quam ad particulare et motum."

14 Albertus Magnus, *De animalibus*, lib. XXI, tr. 1, c. 3, ed. Stadler (1920), 1332: "Et haec est causa quod haec genera animalium similitudines hominis vocantur [...]." On the history of this concept see Janson (1952), 73–89.

15 Albertus Magnus, *De animalibus*, lib. XXI, tr. 1, c. 2, ed. Stadler (1920), 1329: "Pycmeus [sic!] igitur secundum praeinducta quasi medius est inter hominem divinum intellectum habentem et alia muta animalia in quibus nichil divinae lucis esse deprehenditur, in quantum experimentali cognitione utitur per umbram rationis, quam prae ceteris accepit animalibus [...]." According to Janson (1952), 85, this is "the earliest conception of a 'missing link', the earliest attempt to bridge the chasm between mankind and the rest of the animal world." On this see also Köhler (2008), 419f. Salisbury (2011), 122, claims that this idea is not found before the twelfth century but this is wrong as, for instance, Isidor of Seville states that *symia* are endowed with a "*similitudo rationis humanae*," see Isidor of Seville, *Etymologiae*, lib. 12, c. 2, ed. André (1986), 115.

16 See n11 above.

CHAPTER 22

Material Souls and Degrees of Reasoning (John Buridan, Nicole Oresme)

Despite various differences, all of the thinkers we have examined so far – including Albert – agreed that there is at least one crucial difference between human and nonhuman animal souls. In their view, only human beings have rational souls or, more precisely, only humans possess *immaterial* intellectual faculties. For many of them, this lack of immaterial faculties is the main reason why nonhuman animals cannot reason. Even someone like Albertus Magnus, who ascribes certain basic forms of reasoning to nonhuman primates, leaves no doubt that these forms lack many features of intellectual reasoning, most importantly, universals or concepts, for universal cognition and concept formation usually require immaterial faculties (see Part 2). Consequently, those authors struggled with the question of how there can be human-like operations in souls whose nature is quite different from the nature of human souls. But what happens if one questions this difference? What if human and nonhuman souls are very much the same?

In the fourteenth century in particular, these questions were raised and addressed by authors such as John Buridan and Nicole Oresme. Both of them explicitly sympathised with the view that all souls, regardless of whether they belong to a human or a nonhuman animal, are 'material forms' (*formae materiales*). To be clear, neither Buridan nor Oresme declared this to be their own original position. Rather, they noted that it goes back to Alexander of Aphrodisias.[1] Nevertheless, both tried to find arguments in support of Alexander's view.[2] This is not an easy task because the view that the souls of both human and nonhuman animal are material forms gives rise to various puzzles. For instance, in the field of theology, the question is how one can defend the immortality of the human soul if it is, as Buridan puts it, 'extended, deduced from the potency of matter, generated, and corruptible like the soul of a dog or a donkey'.[3] In psychology, this gives rise to the question of how one can

1 On the medieval reception of Alexander see Pluta (1996).

2 The most recent and comprehensive reconstruction of Buridan's arguments is Klein (2016).

3 See, for instance, John Buridan, *Quaestiones de anima* (2nd compilation), Paris, Bibl. Nat., Cod. lat. 15888, f. 70ra, as quoted in Pluta (1996), 95f.: "In hac materia sunt tres opiniones magis famosae. Una fuit Alexandri, quod anima intellectiva humana est forma materialis,

 | DOI 10.1163/9789004363779_028

coherently account for many of the obvious differences that exist between the cognitive powers of humans and nonhuman animals. For even if one says that many nonhuman animals act quite human-like, one cannot deny that humans excel other animals in many respects. But how does one account for many of the obvious differences in cognition if there is no obvious difference with regard to the souls' nature?

According to Buridan and Oresme, there is a relatively simple answer to this question. In their view, the rational souls of humans differ from the souls of other animals with regard to material complexion and complexity. The soul of a donkey is less complex a material form than the soul of a human being.[4] This affects the soul's operations, and so the cognitive operations of a donkey's soul are less complex than those of a human soul. The cognitive capacities of donkeys, in turn, excel those of less developed animals, such as shellfish. Hence, material differences produce all kinds of cognitive differences across the animal kingdom. However, all species differ only 'gradually' (*gradatim*) from each other, as Nicole Oresme points out.[5] In brief, the more complex the soul of an animal, the closer its capacities are to those of the human species. But still the question is, how close do other animals actually come to us? In particular, can they can reason as we can?

For Buridan, at least highly developed animals, such as nonhuman primates (*simia*), 'are capable of reasoning to some extent' (*quodammodo rationabiles*). But dogs and other nonhuman animals also 'reason and syllogise' (*ratiocinantur et syllogizant*), although they reason not 'as subtly and completely' (*ita*

---

extensa et deducta de potentia materiae, generabilis et corruptibilis, ut est anima canis aut asini." For Aquinas, for instance, only nonhuman animals' souls are educed from matter; see Davids (2017), 63–67. On possible sources of Buridan's version of the Alexandrian doctrine see Hoenen (1993).

4 See ibid.: "Et dicebat hoc non debere negari propter magnam subtilitatem hominis vel eius ratiocinationem, quoniam hoc dicebat provenire ex nobilitate complexionis corporis humani vel ex nobilitate animae humanae super alias animas [...]." Actually, it is not entirely correct to say that the human soul is more *complex* because '*complexio*' simply refers to the material composition of the soul. Thus, a difference in complexion does not necessarily imply a difference in complexity. However, Buridan points to the 'nobility' (*nobilitas*) of the human soul which could refer to its complexity.

5 Nicole Oresme, *Expositio et quaestiones in Aristotelis De anima*, lib. III, q. 4, ed. Patar (1995), 335: "Et ideo videmus <quod> aliqua animalia imperfecta et modicae cognitionis nec discurrunt, sicut conchae marinae et talia huiusmodi. Et alia sunt perfectioris cognitionis, et alia adhuc perfectioris, et ita gradatim intantum quod aliqua videntur discurrere et habere notitias similes cognitionibus humanis, sicut sunt simiaeae [sic] aut talia animalia; tamen adhuc quantum adhoc homo superexcedit omnia alia animalia."

subtiliter ac complete) as humans and nonhuman primates.[6] The example he gives in this context is similar to the one presented by John Duns Scotus (see Chapter 19): a dog wants to go to its master, but since there is a pit blocking the road it cannot take the most direct way. Therefore, it takes a detour in order to get to its master even though this way is much longer. While Scotus argues that the dog finds its way by means of a sensory appetite, Buridan claims that the dog 'reasons and syllogises that falling into the pit is not good, therefore [one should make a detour]'. Thus, the dog does not follow its natural instinct but makes some sort of rational decision. Both its behaviour and the cognitive process triggering this behaviour resemble, and may even be the same as, the behaviour and cognition of a human being.

There is obviously at least one serious objection that can be raised against Buridan's theory, namely, the objection that his description of the dog's cognition is unnecessarily anthropomorphic. Even if one agrees that a *human* being would take a detour on the basis of reasoning, one cannot infer from our *human* cognition to the cognition of *nonhuman* beings. Even though dogs and other animals might be clever animals too it seems strange to say the dog's solution to the above-mentioned problem is the result of syllogising. One could even argue that it is incoherent to make such a claim because, as we have seen in Chapter 12, Buridan denies the capacity of concept formation to nonhuman animals. Consequently, it seems inconsistent to claim that the dog forms a premise such as 'Falling into the pit is not good' because this presupposes the capacity to form and arrange concepts such as 'pit', 'good', 'falling', and so forth.

It seems that Buridan would actually agree with this. He would agree that the dog does not actually form premises and conclusions like we do, primarily because it does not possess concepts such as 'pit' or 'good' and so forth. Hence, if we describe or formalise the dog's reasoning in the same way in which we would describe our reasoning, we do indeed anthropomorphise its cognition. But the actual point Buridan wants to make here is that it is similarly flawed to argue that the dog's soul is essentially different from our soul. For how could this be proven? If we try to prove it by referring to the things we can do in

6 See ibid. and John Buridan, *Quaestiones de anima* (2nd compilation), Paris, Bibl. Nat., Cod. lat. 15888, f. 70ra, as quoted in Pluta (1996), 95f.: "[…] sicut dicemus simiam <esse> ingeniosam super cetera animalia et quodammodo esse rationabilem, immo et canes et alia animalia ratiocinantur et syllogizant, quamvis non ita subtiliter ac complete sicut homo vel simia. Quod apparet, quia, si canis videt dominum suum et vult ire ad ipsum et in directa linea inveniat magnam foveam, non intrabit in illam, sed quaerit aliam viam, licet longiorem, quod non faceret, nisi ratiocinaretur et syllogizaret, quod non est bonum cadere in foveam et cetera." On this see also Pluta (2015), 281–283.

comparison to the things dogs can do, we are likely to fail because any difference can be explained by some material difference, as noted above. Therefore, Buridan makes a suggestion which could be summarised as follows: instead of denying the capacity of reasoning to other animals, one could speak of different *degrees of reasoning*. While our reasoning is quite 'subtle and complete', the reasoning of dogs is less so. Nonetheless, the dog is reasoning, and the degree of subtleness or completeness of its reasoning depends on the complexion and complexity of its soul. This also applies to other animals: monkeys, for instance, are even smarter than dogs, whereas shellfish are comparatively dumb.

Like Buridan, Nicole Oresme also states that some of the higher animals 'seem to think and seem to have cognitions similar to human kinds of cognition' (*videntur discurrere et habere notitias similes cognitionibus humanis*). And like Buridan (and Albert), he stresses the particular position of primates (*simiae*).[7] Because of their human-like 'physiognomy' (*physiognomia*), he says, their psychology is very likely to be human-like, too. However, man's exceptional position remains uncontested insofar as only humans possess the capacity to communicate by language. In Oresme's view, this has a huge impact on the capacity of reasoning, because without linguistic vehicles one cannot reason properly. From what he says in this connection, it is not entirely clear whether he actually thinks that nonhuman primates can 'reason' (*ratiocinari*) because he says that 'if they could speak, they would reason like humans' (*si possent loqui, ratiocinarentur sicut homines*).[8]

This passage could, on the one hand, be understood to mean that language is a necessary prerequisite for reasoning. Consequently, creatures without language cannot reason. On the other hand, one could also read Oresme as claiming that if nonhuman primates had language, their reasoning would be as perfect as human reasoning. Nonetheless, they do engage in reasoning even though their reasoning is inferior to human reasoning in the sense that it is a non-linguistic form of reasoning. No matter which reading is to be preferred, it is clear that neither the position of Nicole Oresme nor that of John Buridan comes at the price of giving up man's exceptional position. Man's standing on top of the ladder of beings (or animals, more precisely) remains undisputed.

7 See n5 above.

8 See Nicole Oresme, *Expositio et quaestiones in Aristotelis De anima*, lib. III, q. 4, ed. Patar (1995), 335: "Unde, quia exterior figura quodammodo est signum dispositionis interiorum, ita de physiognomia: inde est quod in quibusdam illa animalia quae magis accedunt ad similitudinem hominis quantum ad figuram sunt maioris industriae, sicut dicebatur de simeis intantum quod videtur aliquibus quod, si possent loqui, ratiocinarentur sicut homines, quamvis tamen in quibusdam ita perfecte."

But unlike the other thinkers covered in this part, they do not consider this special position to depend upon the immateriality of the human soul. Instead, they argue that cognitive superiority is a matter of material complexity. This argument might strike one as problematic since it seems to blur the lines between humans and nonhuman animals. For if we say that other animals reason too we give up human privilege with respect to this power. Reasoning is no longer a peculiar human capacity then, or reasoning is at least no longer what makes humans a peculiar species of animals.

To this Buridan and Oresme would possibly respond that *human* reasoning is still a peculiar form of reasoning for many of the above-mentioned reasons: it is linguistic, more subtle, etc. Furthermore, humans reason about all kinds of matters, practical as well as theoretical, while the reasoning of dogs, nonhuman primates, and other animals is mainly practical and relatively simple. Nonetheless, it counts as a form of reasoning because all reasoning, regardless of whether it takes place in a human or in a nonhuman soul, is 'material reasoning', as one could call it. Reasoning (or proper reasoning at least) is not dependent on the immaterial nature of certain faculties. This theory creates serious theological problems, as has been mentioned, because it could potentially threaten the special status of human beings, especially as regards the immortality of their rational souls. But as far as psychology is concerned, it has an obvious strength: it can easily account for many human-like behaviours in nonhuman animals because most, if not all, kinds of human capacities can, in one way or another, be found in other animals, too.

As we have seen in this part, one can also give alternative explanations for such behaviours. One can, for instance, say that they are brought about by some natural inclination or by sensitive appetite, as Thomas Aquinas and John Duns Scotus claim. Or one can, like Roger Bacon and Albertus Magnus, argue that they are the results of processes taking place in the cogitative power or in some 'shadow of reason'. Regardless of which explanation one prefers, it has become clear that they constitute different ways of dealing with the behaviours of those animals that by definition lack the faculties of intellect and reason. As we have seen, there are at least some late medieval thinkers for whom this lack does not necessarily imply the incapacity to engage in reasoning, despite the differences that exist between the reasoning of human and nonhuman animals. This does not mean that these thinkers present a more coherent account of animal rationality than others. It simply means that they chose a different approach. We shall take up the differences that exist between their approaches in Part 6.

Before that, we will, however, examine another aspect of animal rationality, namely, prudence. Unlike concept formation, judging, and reasoning, prudence is not a part of the triad of intellectual operations. Nonetheless, it gave rise to very similar questions because, on the one hand, prudence seems to require intellect and reason, since acting prudently usually means to act with foresight or planning. But, on the other hand, nonhuman animals exhibit various kinds of prudent behaviours, for example, insofar as they store food for the future. Hence, the question is: Can they be prudent at all and, if so, are they as prudent as we are?

PART 5

Prudence

Introduction to Part 5

The previous parts have examined those cognitive capacities that are part of the triad of intellectual operations, namely, universal cognition or concept formation, judging, and reasoning. All of these operations are intellectual in that they were usually taken to depend on the possession of intellect and reason. Nonhuman animals which, by definition, do not possess these faculties cannot, therefore, engage in any of these operations. However, it has become clear that various questions and doubts can be and have been raised concerning this definition, because it is not always so obvious that nonhuman animals, especially the more highly developed species, do not form concepts, judgments, or syllogisms. The ways in which late medieval thinkers dealt with those doubts and questions differed and the different strategies they choose will be the subject of the next part. But before that we shall look at a fourth cognitive capacity, namely 'prudence' (*prudentia*). This choice is in need of an explanation because prudence does not belong to the triad of intellectual operations. Since it does not, one could ask why prudence is more important or interesting a capacity to look at than, for instance, language or self-awareness. As we shall see, however, prudence is an integral part of late medieval theories of animal rationality. Moreover, it has much in common with the modern notion of *animal intelligence* and so establishes an interesting parallel between past and present theories.

A good example for the role prudence plays in late medieval discussions is Thomas Aquinas' account of reasoning that we have examined in Chapter 19. In contrast to other authors like John Buridan, Nicole Oresme, or Albertus Magnus, Aquinas denies the capacity of reasoning to nonhuman animals. In his view, the dog that has the choice between three different paths only 'quasi-syllogises' when choosing the third way. Prima facie it looks as if the dog reasons, but actually it does not engage in a rational thought process. Rather, its behaviour is driven by natural instinct. The interesting point for the present study is that even though Aquinas does not ascribe reasoning to dogs and other animals, he nevertheless states that they are 'prudent' (*prudentia*).[1] He emphasises that this does not mean that they possess 'reason' (*ratio*) or

1 Thomas Aquinas, *Summa theologiae* I-II, q. 13, a. 2, ad. 3, ed. Leonina VI (1891), 100: "Et ex hoc contingit quod in operibus brutorum animalium apparent quaedam sagacitates, inquantum habent inclinationem naturalem ad quosdam ordinatissimos processus, utpote a summa arte ordinatos. Et propter hoc etiam quaedam animalia dicuntur prudentia vel sagacia, non quod in eis sit aliqua ratio vel electio."

 | DOI 10.1163/9789004363779_029

'choice' (*electio*). 'Reason is fully and perfectly found only in humans', as he clearly says in the *Quaestiones disputatae De veritate*, and this is also the reason why nonhuman animals lack 'free choice' (*liberum arbitrium*).[2] Freely choosing something requires reasoning, insofar as one needs to weigh one option against another and at best chooses the most appropriate or the most rational. Nonetheless, nonhuman animals have what Aquinas calls 'a certain likeness of reason' (*aliqua similitudo rationis*), namely, prudence or 'natural prudence' (*prudentia naturalis*), more precisely. By virtue of prudence they come very close to humans. In a sense, their nature 'touches' (*attingit*) the superior nature of humans.

This kind of argument is somewhat familiar from modern accounts of animal cognition where even those who argue that nonhuman animals are *nonrational*, often agree that they show all kinds of *intelligent* behaviours. This kind of argument seems to do justice to the evolutionary kinship of humans and other animals without actually threatening the animal/human boundary. The problem with this argument, however, is that it remains unclear what it means to say that someone (or something) is intelligent. Are computers intelligent because they run complex programmes? And is my cat intelligent because it steals cheese from my plate when I am not watching? Intelligence thus needs to be defined more clearly. Otherwise, it is just a catch-all term for all sorts of (cognitive) achievements that one does not want to call 'rational' for whatever reason. The same thing applies to prudence because it too seems to be a catch-all term for astonishing behaviours of nonhuman animals. Now Aquinas could, of course, defend the ascription of prudence by referring to Aristotle. In the famous opening passage of the *Metaphysics*, Aristotle states that some animals 'are prudent' (*prudentia sunt*).[3] And everyone who read and commentated on the *Metaphysics* in the Middle Ages was clearly familiar with this statement. However, it remains unclear what it means to say that an animal is prudent.

In fact, there was a relatively distinct definition of prudence in the Middle Ages which largely reflected the ancient notion. John Buridan, for instance, refers to Seneca: 'If you are prudent, like Seneca says, you arrange present

---

2 Thomas Aquinas, *Quaestiones disputatae De veritate*, q. 24, a. 2, co., ed. Leonina XXII.3.1 (1973), 686: "Ratio autem plene et perfecte invenitur solum in homine; unde in eo solum liberum arbitrium plenarie invenitur. Bruta autem habent aliquam similitudinem rationis, in quantum participant quamdam prudentiam naturalem, secundum quod natura inferior attingit aliqualiter ad id quod est naturae superioris [...]."

3 Aristoteles Latinus, *Metaphysica* (*tr. Moerbeka*), lib. I, c. 1, 980b20-23, ed. Guillemin-Diem (1995), 11: "Animalia quidem igitur natura sensum habentia fiunt, ex sensu autem quibusdam quidem ipsorum memoria non infit, quibusdam vero fit. Et propter hoc alia quidem prudentia sunt [...]."

matters, provide for the future, and recall past things'.[4] Similarly, John Duns Scotus says that 'prudence includes foresight concerning future events from memory of past events'.[5] The following example might help to illustrate how this is to be understood: imagine you recently moved to a new neighbourhood. Almost everything is new to you and you are not yet familiar with the local facilities and utilities. One late afternoon you sit at your desk and begin to read a fascinating book. You entirely forget about the time but after several hours you begin to feel hungry. Hence, you stop reading, walk to the kitchen, but all that you find is an empty fridge. You therefore walk to the supermarket but unfortunately it closed fifteen minutes ago. That night you go to bed hungry and promise yourself to be more provident in the future. Next time you carefully check whether the fridge is well-stocked before sitting down to read a book and, if not, you go to the supermarket early. In short, you take measures *in the present* to ensure that what has happened to you *in the past* will not happen again *in the future*. If you manage to do this, you act prudently, according to the classical ancient and medieval definition.

Thus, prudence is a capacity which has at least two different cognitive components: memory and foresight.[6] In modern animal psychology one would say that it includes *mental time travel*. One can mentally go back and forth in time. The question is whether this is the case in nonhuman animals as well.[7] At first glance, it seems to be. The classical example which was used by ancient and medieval writers in this context is the behaviour of ants. Ants store food for the coming winter during summer months, and so they behave prudently because they provide for the future.[8] But do ants really anticipate that the winter is coming? Are they capable of foreseeing a future drought? And do they remember past scarcities of food? In short, are ants endowed with the necessary

4 John Buridan, *Summulae de demonstrationibus* 8.5.4.4, ed. De Rijk (2001), 128: "Nam, sicut dicit Seneca, 'si prudens es, praesentia ordina, futura praevide, praeterita recordare'."

5 John Duns Scotus, *Quaestiones super libros Metaphysicorum Aristotelis*, lib. I, q. 3, §2, eds. Andrews et al. (1997), 87: "[…] prudentia includit providentiam de futuris ex memoria praeteritorum."

6 On the connection between memory, foresight, and prudence see Summers (1987), 266–268.

7 For modern research on this question see Cheke & Clayton (2010).

8 The *locus classicus* for this example is Prov 6:6. Patristic authors often referred to ants as moral role models. See, for instance, Basil of Caesarea, *Homiliae IX in Hexaemeron*, lib. IX, c. 3, ed. Migne (1886), 194C, and Ambrose of Milan, *Hexaemeron*, lib. VI, c. 4, §16, ed. Migne (1882), 262D. It is also mentioned by Isidor of Seville, *Etymologiae*, lib. XII, c. 3, §9, ed. André (1986), 131, as well as in some versions of the *Physiologus*, see tr. Curley (2009), 20–23. On its occurrence in ancient texts, see Dickerman (1911), esp. 124f.

cognitive components of prudence and, if they are, how do these compare to the human capacity to remember the past and anticipate the future? Answering these questions is crucial for determining whether nonhuman animals can be prudent because if they lack the necessary cognitive components, it seems questionable to say that they can act prudently. The lack of intellect and reason plays a decisive role in this context, since the recollection of the past and the anticipation of future events require intellectual powers.

The thinkers that will be covered in this part – Albertus Magnus, Thomas Aquinas, Roger Bacon, Bonaventure, Peter of John Olivi, Giles of Rome, and John Duns Scotus – disagreed on what the lack of intellectual powers amounts to with regard to prudence and its cognitive components. On the one hand, as we shall see, they agreed that many nonhuman animal species are capable of *remembering* but incapable of *recollecting* things. Moreover, most of them held that other animals do not actually anticipate future events. But, on the other hand, they came to different conclusions with regard to prudence: while some of them (e.g. Albertus Magnus and Thomas Aquinas) held that nonhuman animals possess some sort of prudence even though they partly lack the necessary cognitive components, others (e.g. Giles of Rome and John Duns Scotus) argued that to ascribe prudence to other species is to speak metaphorically, that is, prudence cannot be had without intellectual powers. Their discussion thus delivers some interesting insights into the criteria that guided them in ascribing or denying certain cognitive capacities to nonhuman animals.

CHAPTER 23

Memory vs. Recollection (Albertus Magnus)

Memory is an essential component of prudence insofar as acting prudently means to provide for the future by *remembering past events*. Knowing that I must buy bread for tonight's dinner, I try to be at the supermarket early because I *remember* the supermarket's opening hours from my last unsuccessful visit. Remembering something does not seem to be an exclusively human capacity. Indeed, there seems to be no scarcity of examples of memory in nonhuman animals. In Chapter 11, we have already come across Roger Bacon's statement that dogs, monkeys, and many other species distinguish things that are entirely new to them from things they have seen before, that is, 'things of which they have memory' (*res quarum habent memoriam*).[1] Memory is, however, not restricted to mammals or highly developed species: birds and insects return to their nests or houses and they could not do this without remembering where those nests or houses are. According to Augustine, even fish have memory, otherwise they would not swim to the surface of the water when somebody passes by a pond. In his view, they do this because they remember that previous passers-by have pitched food into the water.[2] Observations like these were then given a physiological (or neurobiological) explanation within the theory of inner senses. For instance, Avicenna and Averroes explained that memory is among those inner sense faculties that we share with other animals. According to this theory, most animals are capable of remembering things because there is a certain part of their brain (the rear part, as most thinkers claimed) in which the impressions of things that have been perceived in the past are stored. Both Avicenna and Averroes emphasised, however, that not all animals are capable of actively 'recalling' (*recordare*) these impressions. And so there seems to be a difference between *memory* in terms of recognition and *recollection*, that is, the deliberate search for a past impression.[3]

1 Roger Bacon, *Perspectiva*, pars II, dist. 3, c. 9, ed. Lindberg (1996), 246.

2 See Augustine, *De Genesi ad litteram*, lib. III, c. 8, §12, ed. Migne (1887), 284. On his account of memory in nonhuman animals and its Stoic origins see Brittain (2002), 282–288, and Van Dusen (2014), 184–193.

3 See, for instance, Avicenna Latinus, *Liber de anima seu Sextus de naturalibus*, lib. IV, c. 3, ed. van Riet (1968), 40: "Memoria autem est etiam in aliis animalibus. Sed recordatio quae est ingenium revocandi quod oblitum est, non invenitur, ut puto, nisi in solo homine." On memory and recollection in Avicenna and Averroes see Gätje (1988); Black (1996); Di Martino (2007)

 | DOI 10.1163/9789004363779_030

The Arabic thinkers who developed the theory of inner senses in some detail were not the first to note this difference. It is already present in Aristotle, especially, in his treatise *De memoria et reminiscentia* which became known as a part of the so-called *Parva naturalia* in the Latin West.[4] One of the first to comment on this work was Albertus Magnus who wrote a paraphrase (as second book of *De sensu et sensato*) around the beginning of the second half of the thirteenth century.[5] Of course, this is not the only work in which Albert is concerned with memory but it is the one in which he discusses in the greatest detail the differences that exist between memory (*memoria*) and recollection (*reminiscentia*).[6] As regards memory, Albert agrees with large parts of the tradition, according to which this is a capacity we share with other animals. The evidence for this is, for instance, that many animals return to places where they have been before: birds return to their nests, sheep return to their stables, and vultures return to those spots where they have found a corpse before. They could not do this without memory, because memory is the faculty 'that makes return the sensible impression of something that has been sensed before and is now absent', as Albert puts it in *De animalibus*.[7]

Still, three points need to be noted here. First, not all animals are capable of storing sensory impressions. Flies, for instance, are incapable of remembering the blows we deal out in order to chase them off, as Albert observes. Therefore, they frequently return to us. Furthermore, unlike other animals, they do not have fixed dwellings because they would not remember where those dwellings

---

and (2008), 123–138; Bloch (2007), 145–166. On the reception of their theories in the Latin West see Coleman (1992), 328–362; Di Martino (2006); Müller (2015a); Black (forthcoming).

4 On the reception of Aristotle's theory of memory and recollection in the Latin West see Bloch (2007), 137–228. On medieval theories of memory in general see Yates (1969), 63–113; Carruthers (1992); Coleman (1992), 80–537; Bloch (2014); Müller (2015b). On thirteenth-century theories of memory in nonhuman animals in particular see Köhler (2014), 153–189.

5 A critical Latin edition of this commentary has been published by Silvia Donati as volume 7/2 of the *Editio Coloniensis*.

6 The second important text is the relatively comprehensive collection of questions on memory in *De homine*, eds. Anzulewicz & Söder (2008), 297–312. On Albert's account of memory and recollection see Coleman (1992), 416–421; Anzulewicz (2005); Bloch (2007), 179–195; Müller (2015b), 109–120, and (2017). Albert's views on memory in nonhuman animals are reconstructed by Köhler (2014), 153–189.

7 See Albertus Magnus, *De animalibus*, lib. XXI, tr. 1, c. 2, ed. Stadler (1920), 1326: "[…] memoria est quae facit ex prius per sensum accepto redire in absens sensibile, sicut videmus vultures saturatos recedere a loco cadaveris, et postea iterum redire ex memoria loci et cadaveris: et hoc modo ad caulas revertuntur greges, et aves ad nidos et huiusmodi." See also *De memoria et reminiscentia*, tr. 1, c. 2, ed. Donati (2017), 117.

were or what they looked like.[8] There are also other tiny animals that are not endowed with memory, such as worms and shellfish.[9] The reason for a lack of memory is not, however, the size of an animal. There are many relatively small animals that can store past impressions. A prominent example is bees: they lack the sense of hearing, hence are incapable of being taught because instruction usually requires hearing. Nonetheless, they do have memory; otherwise, they could not find their hives.[10] The reason for an incapacity to store past impressions is the bodily complexion of an animal rather than its size. That is, the organs of flies are too watery and too cold or contain too much soil. This 'immobilises the animal spirit' (*immobilitat spiritus animalis*) which transmits the sensory forms that are perceived through the external senses. Consequently, the capacity of memory is seriously impaired or even totally lacking in such animals.[11]

Whether or not ants belong to this class on Albert's view is unclear. On the one hand, there seems to be no passage in which he explicitly ascribes memory to them. For this reason, some modern interpreters claim that he takes them to lack memory.[12] On the other hand, there is no passage in which he explicitly denies memory to them either. And since he argues that those animals which return to fixed dwellings are endowed with memory, it is likely that ants possess memory too, in his view, because like bees they have nests to which they return.[13]

A second point worth noting concerns the question of whether those animals that store sensory impressions possess an awareness of the temporal aspects that are connected with these impressions. When we remember our last shopping trip, we might also remember what the grocery store looked like: we might recall its colour, its shape, its arrangement, and so forth. When we do this, we are aware of the fact that the image in front of our 'inner eye', so to

8 See Albertus Magnus, *De animalibus*, lib. XXI, tr. 1, c. 2, ed. Stadler (1920), 1326: "Quaecumque autem animalium non persequuntur nisi praesens sensibile et ex prius accepto non revertuntur ad absens sensibile, scimus memoriam nullam prius acceptorum habere, sicut sunt muscae quae cum abiguntur, revolant immemores ictus prius accepti. Videmus etiam quod certum domicilium non custodiunt, et videmus quod non nisi praesens sensibile persequuntur." See also ibid., c. 8, 1345.

9 See Albertus Magnus, *De memoria et reminiscentia*, tr. 1, c. 1, ed. Donati (2017), 115.

10 See Albertus Magnus, *De animalibus*, lib. XXI, tr. 1, c. 2, ed. Stadler (1920), 1327.

11 Albertus Magnus, *Metaphysica*, lib. I, tr. 1, c. 6, ed. Geyer (1960), 8. On this see also Roling (2013), 413f.

12 See, for instance, Anzulewicz (2009), 44.

13 For this argument see Albertus Magnus, *Metaphysica*, lib. I, tr. 1, c. 6, ed. Geyer (1960), 9. On the cognitive capacities of insects in medieval psychology see Guldentops (1999) and Roling (2013b).

speak, is an image of the grocery store we have seen yesterday or a week ago. In short, we know that what we are *currently* recalling is an image of something we have seen *in the past*. The question is, do nonhuman animals also know this? Albert's answer to this question is twofold. On the one hand, he agrees with Aristotle that all animals that are capable of remembering also 'sense time' (*sentiunt tempus*).[14] The reason for this is simply that memory consists in storing *past* perceptions. If a vulture perceives a corpse, stores this perception, and later remembers it, it remembers something from the past, just as we remember something from the past when we recall how the grocery store looked *yesterday*.

There is, however, an important difference between our perception of time and the vulture's, according to Albert. We perceive time 'in itself' (*in se*), whereas the vulture or any other nonhuman animal perceives time 'not in itself' (*non secundum se*).[15] One might say that we have a *direct* perception of time. We know that time is 'a number of motions' (*numerus motus*), as Albert puts it. When recalling an image of the grocery store, for example, we know that a certain number of hours or days have passed since we have seen this store. Nonhuman animals, by contrast, do not know this. Their perception of time is *indirect* or 'obscure', as Albert says, because 'they sense the past through the images of past things that are in them'. This means that they only have perception of time insofar as they store the impressions of things that they have perceived in the past. They certainly notice that what they remember is not currently there; otherwise, sheep would run away frightened just by remembering a wolf. Thus, it is not the case that they do not perceive the past at all. However, they notice the past only by *recognising* something. For instance, they realise that the stable to which they are currently returning is the stable in which they lived 'in the past' (*in praeterito*). Their perception of the past is thus linked to a particular object and its function is to *recognise* this object.

This leads to a third important point, namely, the above-mentioned difference between memory and recollection. To recognise an object means to remember that one has seen this object before. When we stand in front of the

14 Albertus Magnus, *De memoria et reminiscentia*, tr. 1, c. 2, ed. Donati (2017), 116.

15 Ibid., 116f.: "Sentitur enim tempus dupliciter, in se scilicet, secundum quod est numerus motus, et sic rationabilia sola sentiunt et cognoscunt tempus; alio autem modo sentitur tempus in temporali et non secundum se. Et tunc sentitur sub determinata differentia temporis secundum quam adiacet rei temporali. Et hoc est obscure percipere tempus, et sic sentitur in praeterito quando res praeteriit, sicut oves et caprae revertuntur ad caulas, cognoscentes caulas ubi habitaverunt in praeterito [...]. Sentiunt enim praeteritum per picturam praeteritorum quae sunt in ipsis." On this see also Köhler (2014), 207.

grocery store we will *remember* that we have been to *this* grocery store before, and in the same way sheep remember their stable. Hence, they are capable of what Latin authors called '*memorari*'. Yet remembering something is different from *recollecting* something. Recollection or *reminiscentia*, as Albert puts it, with explicit reference to several Greek and Arabic philosophers, is 'the search for something that has been forgotten by memory' (*investigatio obliti per memoriam*).[16] This kind of investigative process 'is not simply the repeated grasp of things perceived before' (*non simpliciter est iterata acceptio prius acceptorum*), as when the image of the grocery store suddenly comes to our mind, fades away, and then comes back again after a while.[17] Rather, recollection is when we have forgotten what the grocery looks like, what its name was, or where it is located and then, by browsing through our memories and by way of association and inference, finally recall the look, name, and address of the store. Obviously, this process is much more complex than standing in front of the store and recognising it. It is actually 'a quasi-syllogistic discourse' (*quasi decursus syllogisticus*) because one jumps from the memory of one thing to another. It is 'not truly a syllogism' (*non tamen est vere syllogismus*), as Albert explains, because it does not involve universals. Nevertheless, its complexity renders it beyond the capacity of nonhuman animals.[18]

One might immediately raise the following objection: recollection is a process that takes place in the faculty of memory, and this faculty is shared by human and nonhuman animals. Hence, there is no reason to suppose that nonhuman animals cannot engage in recollection. In particular, it seems implausible to defend this claim with a view to highly developed nonhuman animals, such as pygmies. Pygmies engage in certain simple syllogistic processes, as we have seen in Chapter 21, and so they should actually be capable of recollection as well.

16 Albertus Magnus, *De memoria et reminiscentia,* tr. 2, c. 1, ed. Donati (2017), 124: "Ponemus igitur primo sententias Averrois et Avicennae et Alexandri et Themistii et Alfarabi, qui omnes concorditer dicunt quod reminiscentia nihil aliud est nisi investigatio obliti per memoriam."

17 See ibid., c. 2, 127.

18 See Albertus Magnus, *De homine*, eds. Anzulewicz & Söder (2008), 302,25: "[…] quaedam inventio exigitur ad reminiscentiam et quasi decursus syllogisticus." And id., *De memoria et reminiscentia,* tr. 2, c. 6, ed. Donati (2017), 135: […] "sed etiam in hoc differunt, quod memoriam participant multa aliorum animalium ab homine, quae tamen potentiam habent cogitativam, sed reminisci nullum, ut ita dicam, participat nisi homo. Causa autem huius est, quod reminisci cum sit investigatio per memoriam, est sicut syllogismus quidam a principio procedens. Non tamen est vere syllogismus, quia procedit per se ex particularibus in res et non in cognitionem aliquam factam per principia […]."

To this objection Albert could give the following reply. Recollection is indeed a bodily process, that is, a process that takes place in a bodily organ.[19] This we know from the fact that we sometimes experience pains when recollecting something: in some moments, we literally rack our brains when trying to recall something. However, recollection is a cognitive process in which the power of memory is assisted by the cogitative power (*virtus cogitativa, virtus distinctiva*) insofar as the cogitative power is responsible for the syllogistic part of the process. Since nonhuman animals, including pygmies, lack this power, in the opinion of Albert, they cannot recollect.[20] Furthermore, recollection requires 'an act of reason' (*actus rationis*), as Albert states in his *De homine* because to recollect something means to search for it deliberately and intentionally.[21] Accordingly, one could say that the power of reason thus primes an act of recollection, and so nobody can engage in recollection without reason.

One might still be unhappy with this answer, because it seems obvious that vultures, for example, recollect the image of a corpse just as we recollect the image of the grocery store. Otherwise, they could not return to the place where they found the corpse a few hours or days ago. They must somehow actively and deliberately recall the image of the corpse and its location in order to get there. This is not, however, how the process works, according to Albert. In his view, the vulture's remembering the corpse is not brought about by recollection, but is triggered by 'natural instinct' (*instinctus naturalis*).[22] This can be understood as follows: when the vulture is hungry, its appetite brings about

19 See Albertus Magnus, *De memoria et reminiscentia*, tr. 2, c. 7, Donati (2017), 135: "Ex his autem quae diximus, quod reminiscentia est cum deliberatione syllogizante, non debet aliquis credere, quod reminiscentia accidat solum secundum intellectum separatum. Cum enim reminiscentiarum finis sit acceptio praeteriti sub metro certo vel incerto, oportet quod sit reminiscentia motus et passio corporea, et reminiscibilitas est secundum dispositiones corporeas complexionales bona vel mala." According to Anzulewicz (2005), esp. 178–182, Albert nonetheless allows for an intellectual memory, especially in his theological works. On the bodily nature of memory and the question of post-mortal memories see Roling (2015).

20 Ibid., c. 1, 124: "Tertium autem est quod Arabes vocant virtutem distinctivam, quae est componens omnia haec et attribuens rei omnibus modis quibus attribui possunt per simile et contrarium et per locum et tempus et aliis modis, ut eliciatur secundum intentum quod cecidit in oblivione, sicut inferius monstrabimus. Et quoad hoc solum reminiscentia solis rationabilibus convenit."

21 See Albertus Magnus, *De homine*, eds. Anzulewicz & Söder (2008), 308f.

22 See Albertus Magnus, *De homine*, eds. Anzulewicz & Söder (2008), 309: "Ad obiectum contra dicendum quod inquisitio non est memoriae, quia inquisitio est actus rationis; sed per metaphoram et similitudinem attribuitur ei, eo quod ipsa conferendo intentiones devenit in imagines et per iudicium imaginum devenit in rem. Et hoc fit in brutis potius per

a desire for food. This then causes to appear the representation of the corpse, since what is stored in the vulture's memory is not only the representation of the dead animal, but also the intention of, say, desirability which moves the appetite.[23] So the difference between us and the vulture is that we can recollect (or deliberately try to remember) the image of the grocery store anytime we want, regardless of whether or not we are hungry. The vulture, by contrast, cannot deliberately recollect a representation of its feeding ground. Instead, the feeling of hunger triggers the recurrence of the image of the food or the feeding ground. As long as it is saturated, it will not remember these.[24] Very likely, this is also the reason why other nonhuman animals, such as apes and pygmies, do not recollect even though they are capable of reasoning to some extent: like any other nonhuman animal they lack the capacity to deliberately start an act of recollection.

In summary, for Albert, human memory differs from the memory of other animals in two crucial respects. First, nonhuman animals have a different perception of time. They do not perceive time in itself, as a certain number of hours or days which have passed. Second, they are incapable of recollecting things. If they remember something, the memory of this object is not brought about by deliberation and a quasi-syllogistic process, but is triggered by other powers of their souls, such as appetite. This is not to say, however, that there are two kinds of memory. If we remember the image of the grocery store, this basically happens in the same faculty as it happens in nonhuman animals. What is different is *how* we come to remember things. Memory in other animals always requires the impulse of another power of the sensory soul and this impulse cannot be controlled voluntarily. By contrast, we do not need to wait for such an impulse but can try to remember whenever, wherever, and whatever we want.

---

modum instinctus naturalis quam per modum deliberationis, ut supra patuit." A similar passage is ibid., 302.

23 From *De homine*, eds. Anzulewicz & Söder (2008), 304, it is clear that both images (*imagines*) and intentions (*intentiones*) are stored in memory.

24 Therefore, it seems a good idea to translate *instinctus naturalis* as 'natural impulse' ("Naturimpuls") in this context, as does Köhler (2014), 177.

CHAPTER 24

Incomplete and Complete Memory (Thomas Aquinas, Roger Bacon)

As we have seen, Albert construes the difference between memory in human and nonhuman animals along the lines of the distinction between memory (*memoria*) and recollection (*reminiscentia*): while humans can both remember and recollect, other animals lack the capacity of recollection. Another version of this distinction is found in Thomas Aquinas and Roger Bacon. They claim that the memory of nonhuman animals is an *incomplete* or *imperfect* memory, whereas humans are endowed with *perfect* or *complete* memory. The question is whether this distinction essentially differs from Albert's distinction. For Albert, there are not different *kinds* of memory: all animals possess memory, no matter whether they are human or nonhuman, but there is a difference in how their memory *works*. Since human memory is assisted by the cogitative power and the faculties of the rational soul, it goes beyond mere recognition. Aquinas and Bacon, in contrast, seem to suggest that there are different kinds of memory, namely complete and incomplete memory.

Bacon discusses the question of whether all animals are endowed with memory in his question-commentaries on Aristotle's *Physics* and *Metaphysics*. In the latter, he argues that 'memory is twofold' (*duplex est memoria*). One type of memory functions 'in connection with sense' (*cum sensu*); the other one goes 'beyond sense' (*post sensum*). Bacon does not explicitly say that the former is sensory and the latter intellectual. But he is clear about the fact that nonhuman animals lack the kind of memory that goes beyond sense. This is the reason why their memory is 'imperfect' or 'incomplete' (*imperfecta* or *incompleta*) in comparison to the 'complete memory' (*memoria completa*) possessed by humans.[1] The difference between the two is that in the case of complete memory there is 'an interruption in time' (*interruptio temporis*), as Bacon puts it. While complete memory can operate apart from the present perception of something, incomplete memory requires 'the presence of a sensible

1 Roger Bacon, *Questiones altere supra libros prime philosophie Aristotelis* (*Metaphysica* I–IV), ed. Steele (1932), 13: "[…] duplex est memoria; quedam est cum sensu, et hec est imperfecta, et hec est communibus animalibus; alia est, que est post sensum, et est cum interruptione temporis, et est completa, et hec est in solis hominibus."

 | DOI 10.1163/9789004363779_031

thing' (*presentia sensibilis*).[2] This sounds like an implausible claim, for why should one speak of memory at all if the object which is perceived is present? Is this not mere sense perception?

One way of making sense of Bacon's statement is to say that what he has in mind here is a classical case of recognition. While we can recall the image of the grocery store at any time and independent of whether we are currently seeing it, a dog, for instance, cannot recall the image of its master unless its master is standing in front of it. Its memory is "simultaneous with sensory perception," as Rega Wood puts it in her interpretation of Bacon's position.[3] When standing in front of its master, the dog can synchronise the image of what it currently sees with the image of what it has seen in the past. If these images match, recognition occurs. This synchronisation does not take place in the case of sensory perception because in this case there is only one image, namely, the image of what is presently seen. In the case of memory, there is one more image, namely, the image of what was seen in the past. This kind of memory is, however, incomplete as it requires the presence of two images, one past and one present. Complete memory does not require two images. For complete memory, it suffices that there is only the image of an object seen in the past. We can, for instance, recall the image of the grocery store *without* presently seeing it, while the dog cannot do this.

If one now compares Bacon's account of memory with Albert's, two questions arise. The first question is whether the incomplete memory Bacon ascribes to nonhuman animals is less complex than the *memoria* Albert attributes to them. The second question is whether complete and incomplete memory are actually two different kinds of memory in the sense of two different faculties. As far as the first question is concerned, it might seem at first as if Bacon's *memoria incompleta* is less complex than Albert's *memoria* because it does not go 'beyond sense', as seen above. That is, the dog cannot actively recall the image of its master. This is, however, a capacity that Albert did not ascribe to nonhuman animals either because to recall something actively means to recollect something, and this requires the assistance of rational faculties, to some extent. As we have seen, this creates certain problems. For instance, how does one explain how vultures return to a corpse? If they cannot recollect the image of the corpse and its location, how can they get there?

---

2 See Roger Bacon, *Questiones super quatuor libros Physicorum* II, ed. Delorme (1928), 125: "[…] dicendum quod hujusmodi animalia memoriam completam non habent, que est per discretionem et deliberationem partium temporis; unde non conferunt presens ad preteritum sicut homo, set solum habent memoriam incompletam, que est cum presentia sensibilis."

3 Wood (2007), 43.

The suggestion was that the appearance of the corpse's image is triggered by another bodily impulse, say, appetite. If the vulture is hungry, the image of the corpse shows up in its faculty of memory and the imagination, and it can then use this image to return to the place where it found the corpse before. This explanation could also be given by Bacon because, like Albert, he endorses Avicenna's theory of inner senses. According to this theory, the faculty of memory is linked to and has certain interactions with other powers of the sensory soul. Hence, even though nonhuman animals cannot recollect, their memory can be activated by the operations of other senses.

From this theory it also follows that incomplete and complete memory are not different faculties. Rather, the power of memory is basically the same in all animals. In contrast to nonhuman animals, however, humans possess an additional operation called 'recollection'. This process does not require a separate faculty (a 'faculty of recollection', so to speak), but takes place in the faculty of memory with the assistance of other powers, especially, the rational faculties. This point is particularly stressed by Aquinas who, like Bacon, ascribes 'imperfect memory' (*memoria imperfecta*) to nonhuman animals.[4] The explanation for this imperfection is that only humans can browse through their memories by some sort of 'reasoning' (*ratiocinatio*).[5] However, memory itself is a power of the sensory part of the soul, and *reminiscentia* is 'a corporeal process' (*corporalis passio*). But 'because of the connection to the intellect', the sensory powers of human beings can engage in such a process. Hence, their memory is 'nobler and more powerful' (*nobilior et virtuosior*) than in other animals.[6] Perhaps one could say that intellect and reason somehow 'upgrade' the sensory powers of humans. Their memory is thus superior to the memory of other animals as it brings about acts of recollection in addition to simple memories or recognition. This does not mean, however, that nonhuman animals entirely lack memory. Rather, their power of memory is limited to remembering, and in this sense it is incomplete or imperfect in comparison to human memory.

---

4 See Thomas Aquinas, *Sentencia libri De sensu et sensato*, tr. 2, c. 1, ed. Leonina XLV.2 (1985), 104.

5 See Thomas Aquinas, *In secundum librum Posteriorum analyticorum Aristoteles expositio*, c. 15, lec. 20, ed. Leonina I (1882), 400b–401a: "[…] et dicit quod cum multa sint talia animalia habentia memoriam, inter ea est quaedam differentia. Nam in quibusdam eorum fit ratiocinatio de his quae remanent in memoria, sicut in hominibus; in quibusdam autem non, sicut in brutis."

6 Thomas Aquinas, *Sentencia libri De sensu et sensato*, tr. 2, c. 8, ed. Leonina XLV.2 (1985), 132f.: "Sic ergo patet quod reminiscencia est corporalis passio nec est actus partis intellective, set sensitive, que in homine est nobilior et virtuosior quam in aliis animalibus, propter coniunctionem ad intellectum."

CHAPTER 25

Foresight and Provision (Albertus Magnus, Bonaventure)

Memory is only one of two cognitive components necessary for prudence. The other essential component is 'foresight' (*providentia*) because to act prudently also means to do something in the present with a view to the *future*. For instance, we go to the grocery store *today* in order to buy things for *tomorrow's* lunch. Of course, we do not strictly speaking *foresee* the future. Rather, we anticipate that we will run out of food soon, and this is why we try to stock up on groceries. Still, it is unlikely that we would do this if we had not some sense of the existence of a future time. We would not store groceries today if we did not suppose that there exists something like tomorrow. Yet, do other animals also have this kind of anticipation of the future?

As mentioned in the introduction to this part, there are many examples of foresightful behaviour in nonhuman animals. Most prominently, ants collect grain in the summer and store it for the winter months. But do they really do this because they anticipate the coming winter? Are they even capable of distinguishing the present from the future? These questions were discussed by several thinkers, and in the following we shall look at the answers of some of them, namely Albertus Magnus, Bonaventure, Thomas Aquinas, Roger Bacon, and Peter of John Olivi.[1]

Albert deals with the question of what was usually called '*providentia*' or the capacity of 'providing for the future' (*providere in futurum*) in various of his writings. Like most other ancient and medieval thinkers, he often refers to the famous example of the ant, but he also devotes some attention to other animals, such as certain species of aquatic animals, which exhibit certain behaviours that seem to be based on this capacity.[2] Octopuses, for instance, frequently ambush other animals, particularly shellfish, as Albert explains in *De animalibus*, and this seems to require foresight. For what the octopus does in order to outsmart its prey is to grab a small piece of rock or a stone, hide itself near the closed shellfish, and then wait for the shellfish to open its shell. Once this happens, the octopus places the stone inside the shell and

1 The views of Albertus Magnus, Thomas Aquinas, and Roger Bacon as well as some other thirteenth-century positions are also discussed by Köhler (2014), 203–214.

2 For a brief summary of Albert's position see Köhler (2014), 206f.

 | DOI 10.1163/9789004363779_032

then sucks out the interior. Similarly, cuttlefish seem to foresee what is going to happen as they tie themselves to rocks before a storm is coming in order to stay safe in troubled waters.[3] In both cases, one can rightly ask whether those animals would be capable of hunting in such a manner or avoiding certain dangers without anticipating future events. The same question could be raised about the behaviour of ants: why would they store food during the summer months if they did not somehow anticipate that there is going to be less food in winter?

On the one hand, Albert admits that such behaviours seem to be based on foresight. Ants seem to store food because they are driven by 'an anticipation of and worry about future scarcities' (*praemeditatio et sollicitudo futurorum defectuum*). On the other hand, he points to the consequences of this explanation: if one accounts for such behaviours by foresight, one must ascribe some 'light of reason' (*lumen rationis*) to such animals because the anticipation of future events is usually considered to require the power of reason. This is, however, clearly opposed to the view that nonhuman animals lack reason. Surely, as Albert points out, one could argue like Pythagoras and claim that there is the same kind of soul in all animals. Within this theory there is no problem with the explanation of similar behaviours in different species, because they all share the same set of cognitive capacities.[4] But how do we then explain the obvious differences that exist? How can one explain that many nonhuman animals, as smart as they might be, nevertheless differ from humans in many regards? There are many species of nonhuman animals whose members always do things 'in the same way' (*uno modo*): all ants store grain in the same way and swallows built their nests in the same way. Humans, by contrast, often

3 See Albertus Magnus, *De animalibus*, lib. XXI, tr. 1, c. 6, ed. Stadler (1920), 1337f., for the example of the octopus and ibid., 1340, for the example of the cuttlefish. Similar stories are told by Basil of Caesarea, *Homiliae in Hexaemeron* VII.3 and VII.5, ed. Migne (1886), 154A-B and 159A.

4 See Albertus Magnus, *De animalibus*, lib. VIII, tr. 6, c. 2, ed. Stadler (1916), 672: "Adhuc autem opera providentiae inveniuntur in quibusdam, sicut formicae et apes thesaurizant cibos: et hoc non videtur posse fieri nisi cum praemeditatione et sollicitudine futurorum defectuum: et constat quod talis praemeditatio non est sine aliquo rationis lumine, quod secundum praecedentis bruta non habent. Videtur autem etiam valde difficile de hiis invenire rationem, quoniam cuius est actus et operatio, illius est potentia eiusdem actus et operationis, et cum operationes insint, videbitur difficile esse reddere modum, quo modo non insint brutis potentiae operationum illarum. Et propter haec forte et hiis similia Pythagoras omnibus unam sive unius rationis animam dixit inesse, et unius rationis esse animam bovis et hominis et aliorum animalium."

differ in how they do things.[5] These differences can hardly be explained by Pythagoras' theory.

Albert, therefore, tries to come up with a theory that can account for the foresightful behaviour of nonhuman animals without ignoring the differences that exist between different species, in particular between human and nonhuman species. According to this theory, ants store food not because they make some 'conjecture about the future' (*coniecturatio futuri*). Rather, they are driven 'by the avidity for the food that is presently there' (*ex aviditate cibi praesentis*).[6] In other words, there is some sort of innate mechanism in an ant that makes it collect food. Certainly, its 'present acts are directed at future times' (*actus praesentes ordinati sunt ad futura*) because it collects food today and eats it tomorrow, but it does not have what Albert calls a 'cognition of future states' (*cognitio futurorum*). Unlike humans, who buy bread *knowing* that they are going to run out of bread in the near future, the ant collects and stores grains *without* being cognitively aware of a future lack of food. It is incapable of imagining a future period or a future state, and so its behaviour is based on 'the imagination of present things' (*imaginatio praesentium*).[7]

Although this theory can account for the ants' behaviour without jeopardising the animal/human boundary, one might still be dissatisfied. First, one could object that it does not explain why there is more collecting activity during the summer than during the winter months. Second, one might have doubts that it can account for the foresightful behaviours of other animals, such as the octopuses' hunting behaviour. Albert is well aware of these problems and provides the following reply. Although it seems as if various animals anticipate future events and thus possess the capacity of foresight, it is, in fact, some *present* stimulus which triggers a particular reaction. Ants perceive an increase in humidity or a decrease in temperature, and this immediately

---

5 Albert points to this fact in *De anima*, lib. III, tr. 1, c. 2, ed. Stroick (1968), 168. On this argument see also Di Martino (2008), 109, and Anzulewicz (2009), 44.

6 See Albertus Magnus, *De animalibus*, lib. VIII, tr. 6, c. 2, ed. Stadler (1916), 673: "Providentiae autem opera quaedam eorum participant sine omne praemeditatione futurorum, sed naturae instinctu: et quando congregant, non coniecturant futurum temporis defectum, sed ex aviditate cibi praesentis: propter quod etiam quaedam animalia sicut apes plus multo congregant quam sufficiat ad nutrimentum per unam hyemen: et hoc faciunt, ut dictum est, ex cibi praesentis aviditate sine futuri coniecturatione." See also Anzulewicz (2009), 49f.

7 Albertus Magnus, *Quaestiones super De animalibus*, lib. VIII, q. 23, ed. Geyer (1955), 196: "Dicendum, quod formica grana colligit. [...] Nec tamen grana colligit, quia cognitionem futurorum habeat, sed actus praesentes ordinati sunt ad futura, et ideo ex imaginatione praesentium faciunt aliqua, quae ordinantur ad futura."

triggers the reaction of collecting grain.[8] Similarly, cuttlefish respond to the presence of certain vapours that occur before a storm and octopuses respond to the present movements of shellfish.[9] Thus, according to Albert, the foresightful behaviour or 'provision' (*provisio*) which is found in many nonhuman animals is based on 'the sensory perception of something that is nearby and either useful or harmful' (*a sensu conferentia aut nocentis vicini*) rather than on 'a precognition by the estimative power' (*praecognitio aestimationis*).[10] This explanation is concise insofar as it can account for the foresightful behaviour of nonhuman animals without actually attributing foresight to them.

As Bonaventure points out, it is also concise in another respect, namely, with respect to free choice. Foresightful actions usually originate 'from an act of deliberation and precognition' (*ex actu deliberationis et praecognitionis*). In other words, when we act with foresight, we deliberate on what would be the best thing to do in order to achieve a certain goal.[11] For instance, we decide that going to the grocery store now is better than going there later because it will soon close. Consequently, attributing foresight to nonhuman animals means attributing free choice too. And unless one changed the premise according to which they lack intellect and reason, this would also mean detaching free choice from the possession of rational faculties. A theory which explains the foresightful behaviour of nonhuman animals by referring to natural powers – such as 'natural instinct' (*instinctus naturalis*) and 'natural sagacity and

8 See Albertus Magnus, *De memoria et reminiscentia*, tr. 1, c. 2, ed. Donati (2017), 117: "[…] formicae colliguntur in domibus suis praesentientes pluvias in futuro. Sed tamen minus sentiunt bruta futurum quam praeteritum. Sentiunt enim praeteritum per picturam praeteritorum quae sunt in ipsis. Futura vero non praesentiunt nisi signo aliquo praesenti, sicut vapore vel calore vel aliquo huiusmodi."

9 See Albertus Magnus, *De animalibus*, lib. XXI, tr. 1, c. 6, ed. Stadler (1920), 1340.

10 Ibid.: "[…] istae tamen providentiae sunt omnium animalium et causantur a sensu conferentis aut nocivi potius quam a praecognitione aestimationis. […] Provisio autem quae est a sensu conferentia aut nocentis vicini, non negatur alicui animali. Omnia enim visu vel auditu aut certe odoratu aut gustu aut tactu nociva vicina fugiunt et conferentia vicina persequuntur."

11 See Bonaventure, *Commentarius in secundum librum Sententiarum*, d. 25, pars 1, a. unicus, q. 1, ed. Collegium S. Bonaventurae (1885), 594: "Ad illud quod obiicitur, quod liberi arbitrii est providere; dicendum quod providentia quaedam provenit ex actu deliberationis et praecognitionis, quaedam provenit ex instinctu naturali. Prima est libertatis arbitrariae et reperitur in solis rationalibus; secunda vero est naturalis sagacitatis et industriae; natura enim prudentissima est, et haec reperitur in brutis animalibus, et talis non ponit liberum arbitrium."

industry' (*naturalis sagacitas et industria*) – can, however, help us to avoid this consequence.[12]

To be clear, the connection between foresight, reason, and free choice is not entirely dissolved. Rather, the argument is that the individual animal lacks the capacity for foresight. Yet nature – that is, the divine creator – acts with foresight, as Bonaventure says, because nature has given this sagacity to the animal. This argument is in fact very similar to the argument made by Thomas Aquinas in his discussion of the capacity of reasoning (see Chapter 19): nonhuman animals do not reason but only 'quasi-reason', he argued. But they have been created or designed by a rational being, namely, God. As the next chapter on his discussion of foresight shows, this argument can also be applied to this capacity.

12 For Roling (2011), 239, n. 62, this natural sagacity is the 'counterpart' to human foresight because it excludes the use of reason. He also points to the similarity between Bonaventure's and Giles of Rome's position.

CHAPTER 26

Quasi-Foresight and Quasi-Hope (Thomas Aquinas)

Like Bonaventure, Aquinas also begins with the assumption that acts of foresight are usually based on deliberation and free choice. Thus, they are distinct from the preventive behaviour of nonhuman animals because they are based on a 'judgment of the intellect' (*iudicium intellectus*). In order to illustrate this point, Aquinas gives the example of a person who has a fever but feels the desire to drink a glass of white wine. In this case, her intellect will judge that it is better to stay away from wine because the consumption of an alcoholic beverage would aggravate the fever. Consequently, she will not give in to her desire, because she is capable of anticipating its bad consequences.[1] Other animals do not have this capacity. If they have a bodily desire, there will be no intellectual judgment preventing them from doing whatever they desire. Surely, they do have a sense of what is good or bad for them. They cognise the '*ratio convenientis et nocivi*', as Aquinas puts it. This means that they sense whether something is, for instance, harmful and thus to be avoided. Yet, unlike our knowledge about the harmfulness of things which is based on intellectual judgments and 'rational enquiry' (*inquisitio rationis*), their cognition of harmfulness derives from estimative judgments which are nothing but instinctual responses to certain stimuli, as we have seen in Chapter 15.[2] Therefore, they might do something without anticipating its consequences. A sheep might, for instance, eat too much grass because it has no intellect which tells it that eating too much grass is likely to cause colic.

Besides intellectual judgments, nonhuman animals also lack what Aquinas calls a 'notion of the future' (*ratio futuri*).[3] This means that even if certain

---

1 See Thomas Aquinas, *Sentencia libri De anima*, lib. III, c. 9, ed. Leonina XLV.1 (1984), 246: "[…] non solum cognoscunt quod in presenti est, set considerant preteritum et futurum, quia intellectus quandoque ab aliquo concupiscibili retrahere iubet, propter considerationem futuri, sicut cum febrienti, ex iudicio intellectus videtur esse abstinendum a vino ne febris incalescat."

2 See Thomas Aquinas, *Scriptum super Sententiis* III, d. 26, q. 1, a. 1, ad 4, ed. Moos (1933), 815: "[…] sicut animalia cognoscunt rationem convenientis et nocivi non per inquisitionem rationis, ut homo, sed per instinctum naturae qui dicitur aestimatio; ita etiam cognoscunt aliquid futurum est, sine hoc quod cognoscant rationem futuri, non conferendo praesens ad futurum, sed ex instinctu naturali, secundum quod aguntur ad aliquid agendum vel ex impulsu naturae interioris vel exterioris."

3 On Aquinas' position see also Köhler (2014), 208–210.

 | DOI 10.1163/9789004363779_033

animals, such as ants, seem to anticipate certain events or provide for the future this is not because they have 'an imagination of the future' (*ymaginatio futuri*). Instead, such animals show a certain foresightful behaviour because 'they imagine present acts which are directed at a certain goal by virtue of natural instinct rather than by an apprehension' of this goal.[4] This lack of apprehension is a crucial point which needs to be spelled out more clearly. Aquinas admits that it seems implausible to claim that animals do not somehow apprehend future states, because they often seem to anticipate future events and humans even observe their behaviour for predicting the future. Sailors, for example, watch the movements of dolphins because when dolphins come to the surface of the water, this is a sign of approaching storms. Similarly, rain is likely to come soon if ants are seen to be storing as much grain as they can in their dry houses.[5] This is also the case when crows cry more frequently than usual.[6] Thus, it seems as if various animal species clearly know what is going to happen. But how is this possible if they only grasp present things?

To this question Aquinas gives a reply which is similar to the one given by Albertus Magnus. In his opinion, dolphins, ants, or crows do not actually *foresee* future events but simply *react to present stimuli*. Such stimuli not only cause these animals' behaviour, but also cause those things they seem to foresee. For instance, an increase in humidity causes a storm, and even as it is doing so it makes the ants immediately store the grains they have collected.[7] Because of this double causation, as one could call it, the observation of animals' behaviour can in fact provide a basis for predictions about future weather conditions, Aquinas argues, even though the animals themselves do not actually anticipate future events.[8]

---

4 Thomas Aquinas, *Sentencia libri De anima*, lib. II, c. 29, ed. Leonina XLV.1 (1984), 194: "quod autem operantur propter finem quasi providentes in futurum, non contingit ex hoc quod habeant aliquam ymaginationem illius futuri, set ymaginantur actus presentes qui ordinantur ad finem ex naturali inclinatione magis quam ex apprehensione."

5 See Thomas Aquinas, *Scriptum super Sententiis* III, d. 26, q. 1, a. 1, ad 4, ed. Moos (1933), 815: "Unde ex eorum operibus homines possunt aliquid scire de hujusmodi futuris, sicut nautae praesciunt tempestatem futuram ex motu delphinorum ad superficiem aquae ascendentium; et formicae veniente pluvia reponunt granum in cavernis." The example of dolphins is presumably inspired by Isidor of Seville, *Etymologiae* XII.6.11, ed. André (1986), 189.

6 See Thomas Aquinas, *Summa theologiae* II-II, q. 95, a. 7, ad 2, ed. Leonina IX (1897), 325. As Hünemörder (1988), 203, observes, Aquinas takes this example from Isidor too; see *Etymologiae* XII.7.44, ed. André (1986), 259.

7 The idea of a natural sensitivity to changing weather conditions is also found in earlier authors such as Basil of Caesarea, *Homiliae IX in Hexaemeron* IX.3, ed. Migne (1886), 194B–195A.

8 See Thomas Aquinas, *Summa theologiae* II-II, q. 95, a. 7, co., ed. Leonina IX (1897), 325: "Respondeo dicendum quod motus vel garritus avium, vel quaecumque dispositiones huiusmodi

To this argument one might object that our observation of such behaviours does not involve foresight either. We are as incapable of foreseeing the future as other animals since all we do is to interpret certain signs *in the present*. We observe the ants *now*, and from this we infer what is likely going to happen *later*. This inference is, however, based on *past* observations. That is, in the past we have observed a correlation – not a causal relation! – between the behaviour of ants and changes in the weather pattern. And so we do not actually foresee the future, but only make predictions based on past observations.

Aquinas completely agrees with this point. Nonetheless, he stresses that making predictions requires the capacity of distinguishing between past, present, and future states. In his view, nonhuman animals lack this capacity. When ants rush to store their grains, they do not do this because they have noticed that there is a correlation between increasing humidity and rainfalls. Rather, they simply react to the present increase in humidity without making any predictions about the future. Unlike us, they cannot interpret present phenomena as signs of future events, mainly because they have no notion of the future. However, there are some cases in which nonhuman animals seem to make predictions, as Aquinas concedes.

In a passage from the *Summa theologiae* he gives the example of a dog that tries to catch a hare.[9] As long as the hare is far away, the dog does not try to catch it, Aquinas says, because it has no hope of doing so. But once the hare starts coming closer, the dog attacks because now it hopes to catch it. Since hope is always directed at a *future* good, this example indeed seems to show that the dog does something *in the present* in order to achieve something *in the future*. For Aquinas, however, the dog's hope is only 'quasi-hope', as one could call it, because, in his view, the dog starts its attack on the hare only '*as if* in the hope of catching it' (*quasi sub spe adipiscendi*). So, just as the dog's reasoning about the escape of the deer is only 'quasi-reasoning', the dog's hope of catching the hare is only 'quasi-hope'. But this gives rise to the question of why the

in rebus consideratae, manifestum est quod non sunt causa futurorum eventuum: unde ex eis futura cognosci non possunt sicut ex causis. Relinquitur ergo quod si ex eis aliqua futura cognoscantur, hoc erit inquantum sunt effectus aliquarum causarum quae etiam sunt causantes vel praecognoscentes futuros eventus."

9 See Thomas Aquinas, *Summa theologiae* I-II, q. 40, a. 3, co., ed. Leonina VI (1891), 267: "[...] dicendum quod interiores passiones animalium ex exterioribus motibus deprehendi possunt. Ex quibus apparet quod in animalibus brutis est spes. Si enim canis videat leporem, aut accipiter avem, nimis distantem, non movetur ad ipsam, quasi non sperans se eam posse adipisci: si autem sit in propinquo, movetur, quasi sub spe adipiscendi." This passage is also analysed by Hünemörder (1988), 203f.; Roberts (1992), 294; Loughlin (2001), 52–54; Miner (2009), 218f.; Perler (2011), 70f., and (2012a), 81; Köhler (2014), 214.

dog does not really hope. Why should it attack the hare without the actual hope of catching it?

In Aquinas' opinion, there are two main reasons for why the dog does not actually hope for something. The first is that the dog does not apprehend the point of time at which it might catch the hare as a *future* point of time. It behaves only '*as if* it foresaw the future' (*ac si futurum praevidet*), according to Aquinas. Prima facie, this seems to be an incoherent claim because it cannot explain why the dog does not always attack the hare. But, for Aquinas, there is a simple explanation: instead of hope and foresight it is 'natural instinct' (*instinctus naturalis*) which moves the dog. This can be understood as follows: when the dog perceives that the hare is far away, no offensive reaction is triggered. But when the hare comes closer, the dog reacts by launching an attack. It does not make any predictions about the hare's movements nor does it hope to catch the hare. Rather, its senses simply respond to what they presently perceive. One might compare the dog's senses to a motion detector. As long as no object appears within the detector's radius, no spotlight is turned on. Similarly, no offensive reaction is triggered in the dog as long as the hare moves outside a certain radius.

In Chapter 19, we discussed the strengths and weaknesses of such a comparison. One of the weaknesses mentioned was that it seems to reduce nonhuman animals to soulless machines. This is misleading insofar as Aquinas clearly considers all animals to be ensouled beings, while he does not attribute souls to artificial beings. Still, there is an important parallel between the two: just as a machine, say, a motion detector, has been designed and programmed by an intelligent, rational being, say, a human engineer, the dog has been 'designed' and 'programmed' by an intelligent, rational being as well, namely the 'divine intellect' (*intellectus divinus*).[10]

This is the second reason why the dog's behaviour is not based on actual foresight and hope: while our will, that is, the 'appetite of an intellectual nature' (*appetitus naturae intellectivae*), 'is moved by an apprehension of the intellect to which it is linked' (*movetur ex apprehensione intellectus coniuncti*), the sensitive appetite of nonhuman animals 'follows the apprehension of the separate intellect' (*sequitur apprehensionem intellectus separati*).[11] This does

10 See Thomas Aquinas, *Summa theologiae* I-II, q. 40, a. 3, co., ed. Leonina VI (1891), 267: "[…] quamvis bruta animalia non cognoscant futurum, tamen ex instinctu naturali movetur animal ad aliquid in futurum, ac si futurum praevideret. Hujusmodi enim instinctus est eis inditus ab intellectu divino praevidente futura."

11 Ibid.: "Ut enim supra dictum est, appetitus sensitivus brutorum animalium, et etiam appetitus naturalis rerum insensibilium, sequuntur apprehensionem alicuius intellectus,

not mean that the dog lacks cognition; it definitely sees, hears, or smells the hare and its perceptions are necessary for triggering its reaction. Still, this reaction is not the result of rational deliberation or free choice.[12]

Yet even if one accepts that the dog's behaviour is not based on deliberation one might still say that its behaviour involves foresight insofar as 'its appetite is moved by something future' (*movetur eius appetitus in aliquod futurum*), as Aquinas puts it.[13] This is true, but only in the sense that appetites or 'present acts are *directed towards* a goal' (*actus presentes ordinantur ad finem*).[14] That is, the dog's attack is directed towards the goal of catching the hare, but the dog itself is neither cognitively aware of the fact that it has this goal nor does it grasp that a goal is something that, hopefully, will be achieved in the future. Thus, the dog's behaviour is not, strictly speaking, *goal-directed*, to use Fred Dretske's terminology. In the opinion of Dretske, there is an important difference between behaviour that is directed *towards* a goal and behaviour that is directed *by* a goal.[15] In order to illustrate this difference, he refers to Davidson's famous example of a heat-seeking missile which moves towards a particular target.

According to Davidson, an "uninformed observer might be justified in attributing a desire and beliefs to the missile; but he would be wrong."[16] He would be wrong, Davidson explains, because the missile "moves as it does because it was designed and built by people who had the very desire and beliefs" that it move in such a way. The missile itself has neither desires nor beliefs, and everybody would agree that if people die from its impact, it is not the missile which is to blame but those people who constructed and launched it, because the missile is not an intentional being. Thus, it "may be driven toward a goal by

---

sicut et appetitus naturae intellectivae, qui dicitur voluntas. Sed in hoc est differentia, quod voluntas movetur ex apprehensione intellectus coniuncti: sed motus appetitus naturalis sequitur apprehensionem intellectus separati, qui naturam instituit; et similiter appetitus sensitivus brutorum animalium, quae etiam quodam instinctu naturali agunt."

12 According to Perler (2012a), 81, n. 24, Miner claims that Aquinas attributes hope to animals absolutely. This, however, is not correct because, on the one hand, Miner (2009), 218f., rightly points out that "[i]n the case of hope, there is no explicit terminological differentiation" between the hope of humans and nonhuman animals. On the other hand, he stresses that "for human hope to be rational, the apprehension of something as possible or impossible requires deliberation."

13 Thomas Aquinas, *Summa theologiae* I-II, q. 40, a. 3, co., ed. Leonina VI (1891), 267: "[...] tamen ex his quae videt animal in praesenti, movetur eius appetitus in aliquod futurum vel prosequendum vel vitandum."

14 See n4 above.

15 See Dretske (1988), 124–127.

16 Davidson (1982), 323.

certain cybernetic mechanisms, but it is not goal-directed," as Dretske puts it, because it is entirely unaware of the fact that it has a goal.[17] Again, one needs to be careful in comparing a dog to a missile. As mentioned before, there are crucial differences between living beings and artefacts, according to Aquinas. But as regards the cognitive awareness for the existence of goals and future states they do not differ that much because neither a dog nor a missile knows that it is moved by a certain goal, which is why, in comparison to humans, a dog only has 'quasi-hope' and 'quasi-foresight'.

17 Dretske (1988), 126.

CHAPTER 27

Operating for and towards the Future (Roger Bacon, Peter of John Olivi)

The question of whether certain nonhuman animals' behaviours are goal-directed in some sense was discussed and spelled out even more clearly by Roger Bacon and Peter of John Olivi. In his commentary on four of the books of Aristotle's *Physics*, Bacon remarks that foresight is no trivial topic, because it is intimately linked to the question of rationality. The reason is that 'the cognition of future things' (*cognitio futurorum*) is one of the most eminent cognitive operations. Therefore, it is likely to require eminent cognitive powers: intellect and reason.[1] But how shall one deal with the foresightful behaviours of nonhuman animals? Birds build nests and ants store grain, and so they seem to 'anticipate' (*precognoscunt*) the coming season. If this anticipation depends upon intellect and reason, it follows that they are endowed with such faculties.[2]

The problem with such a conclusion is that it obviously runs counter to the common definition of the animal/human boundary. So one must either revise this definition or come up with an alternative explanation for foresightful behaviour in nonhuman animals. Since there are many arguments and authorities which validate the common definition of the animal/human boundary, the revision of this definition is not an option, according to Bacon.[3] Instead, he suggests drawing a distinction between three kinds of cognition of future things.[4]

First, there is God's cognition of future things. As God is the first cause of everything, he knows present, past, and future things alike by some sort of

1 See Roger Bacon, *Questiones supra libros quatuor Physicorum* II, ed. Delorme (1928), 121: "Major etiam sic potest confirmari; nobilissima cognitio a nobiliori virtute emanat; set nobilissima cognitio est futurorum, quia difficilius, subtilius, et magis arduum est cognoscere futura quam alia, ergo cognitio futurorum ab intellectu erit."

2 See ibid.: "[...] set hujusmodi animalia per cognitionem temporis futuri hujusmodi operationes faciunt, ergo hujusmodi animalia suas operationes ab intellectu vel ratione faciunt. Minor patet; ad quid enim aves facerent nidos nisi precognoscerent estatem futuram? In hyeme enim pulli vivere non possunt; similiter formica grana colligit in estate, ex quibus in hyeme quam precognoscit futuram nutriatur."

3 See ibid., 122.

4 See Roger Bacon, *Questiones supra libros quatuor Physicorum* II, ed. Delorme (1928), 123f. On this distinction see also Köhler (2014), 211.

 | DOI 10.1163/9789004363779_034

'introspection' (*inspectus*). Second, there is man's cognition of future things, and this is based on 'industry, deliberation, and perspicacity' (*industria, deliberatio, et perspicacitas*). For instance, if astronomers make predictions about future events they infer many things from a close observation of the stars which influence the earthly sphere. Both kinds of cognition require intellect and reason.

Yet, there is also a third kind of cognition of future things, and this is the sort of cognition of the future that is found in many nonhuman animals. Unlike God or human beings, they do not cognise future things by virtue of intellect and reason, but 'by natural instinct or industry alone' (*solo nature instinctu, vel industria*).[5] But what does this instinctual cognition amount to? Bacon here draws a distinction that is very similar to the one Fred Dretske employs above. In his view, humans 'act *towards and for* the future' (*agunt ad futurum pro futuro*), and this is what we mean when we say that they 'provide' (*providunt*) for the future. Nonhuman animals, by contrast, do not act with foresight because 'they only operate *towards* the future and *not for* the future' (*solum ad futurum et non pro futuro operantur*). This means that they do things from which they will profit in the future but they do this without discerning the gap between the present and the future.[6] This argument, however, might occasion some doubts. For why would ants collect grains during the summer and store them for the winter if they did not conceive of the present and the future 'as distinct [times]' (*ut diversa*)? They would hardly do so, it seems, if they did not understand that what they collect *now* is to be kept for *later*. Therefore, they must be capable of discerning different periods of time, and thus act with foresight.[7]

---

5 See ibid., 124: "Tertio modo est cognitio futurorum solo nature instinctu, vel industria et hec cognitio est in animalibus brutis, de quibus loquimur hic, ad regimen vite sue et conservationem, et propter sui in sua simili specie continuationem; et hujusmodi futurorum cognitio non est ab intellectu vel ratione, ut visum est." Bacon makes a similar point in the *Questiones altere supra libros prime philosophie Aristotelis* I, ed. Steele (1932), 16f. See also Wood (2007), 43f.

6 See Roger Bacon, *Questiones supra libros quatuor Physicorum* II, ed. Delorme (1928), 124: "Ad cujus intelligentiam notandum quod quedam animalia agunt ad futurum pro futuro, ut homines, et hujusmodi animalia provident; alia sunt que solum ad futurum et non pro futuro operantur, ut animalia bruta, de quibus loquimur, et istis non attribuitur providere. Et nota quod agere ad futurum non pro futuro, est agere aliquid in presenti vel facere quo agens in futuro indiget vel indigebit, non tamen futurum precognoscit nec differentias temporis discernit, et sic agunt animalia bruta."

7 See ibid.: "Set contra hoc potest sic obici: hujusmodi animalia bruta ut formice aut accipiunt presens et futurum ut unum, aut ut diversa. Non ut unum, quia sic non colligerent grana in estate nec in loco ad conservandum reponerent, cum non indigerent, si estatem presentem

Instead of rejecting this objection out of hand, Bacon tries to take it seriously and makes the following suggestion. In fact, nonhuman animals discern between the present and the future, but they do not do this 'in an essential way or essentially' (*modo essentiali vel essentialiter*) but only 'in an accidental way or accidentally' (*accidentali vel accidentaliter*).[8] In order to illustrate this difference, he gives the following example: the vegetative soul is responsible for nourishing as well as providing for the generation of a living being. Yet, by taking care of generation it not only guarantees the survival of an individual being, but also secures the continuation of an entire species. This, however, happens accidentally. That is, the vegetative soul does not essentially intend to take care of the continuation of living beings in their entirety. Similarly, ants try to survive, but without essentially cognising that what they presently collect will help them to survive in the future: 'their cognition and intention are accidental' (*cognitio et intentioni eorum est accidens*), Bacon says. The ants do not collect grain in the present because they reckon that they are likely to run out of grain in the future. And so even though they seem to care for the future, they actually only care for the present.

The same point was stressed by Peter of John Olivi. According to Olivi, many animals obviously 'aim at future things' (*futura indendant*), for instance, when they desire to eat something to get rid of hunger. They are hungry *now* and wish to be full *later*, which means that they desire to be in a certain state in the near future. Still, they 'do not fully discern' (*non plene discernantur*) between present and future states, as Olivi makes clear. Rather, they treat future states

et hyemem futuram pro eodem acciperent, ergo manifestum est quod presens et futurum accipiunt ut diversa; set sic non possunt accipere nisi distinguendo et discernendo unum ab altero, et si sic, ergo providere et deliberare possunt."

8 Ibid., 125: "Ad hoc dicendum quod hujusmodi animalia bruta presens et futurum accipiunt ut diversa, modo tamen accidentali vel accidentaliter, et non modo essentiali vel essentialiter; set sic accipere ad providendum et deliberandum non sufficiunt, quod per hoc exemplum potest fieri manifestum: anime vegetative duo sunt opera, scilicet nutrire et generare; vegetativa enim decindit semen per istam autem decisionem fit continuatio esse universalis; set vegetativa hoc non intendebat essentialiter, set solam generationem: quod autem ex generatione hujusmodi continuatio accidat, hoc intentioni vegetative accidit. Similiter formice in presenti estate grana colligunt sui conservationem intendentes essentialiter, et non tamen quia hyems sit futura cognoscunt: unde sicut in estate propter conservationem presentem nutrimento utuntur et grana comedunt, sic similiter illam collectionem granorum propter presentem conservationem faciunt; nisi enim hoc facerent, sui statim corruptionem fieri estimarent; et ita si ex hujusmodi collectione granorum accidit conservatio et regimen sue vite in futuro, hoc cognitioni et intentioni eorum est accidens, et ita presens et futurum essentialiter accipiunt ut enim unum, tamen accidentaliter ut duo."

or things 'as if they were presently correlated and connected to those things which they presently sense and do' (*sunt quasi praesentialiter correlata et connexa iis quae praesentialiter sentiunt et agunt*).[9] To give an example, a dog that is used to getting a bone for dinner, is capable of thinking of this bone *before* dinner time. However, when it imagines the bone, this act of imagination does not include the information that the bone is something it is going to receive *in the future*. In this respect, the dog's cognition differs from the cognition of its master, because its master is well aware of the difference between the present and the future. Of course, the dog will not go for a bone where there is no bone. But when there is a bone it will try to get it and it will not waste time reflecting on the fact that it is going to eat the bone *in a second*. Thus, Olivi agrees with the other philosophers that there is a difference between the way in which humans and nonhuman animals perceive time. Only the latter can clearly distinguish between the present and the future, for example, and so only humans can act with foresight in the true sense of the term.

9 See Peter of John Olivi, *Quaestiones in secundum librum Sententiarum*, q. 155, ed. Jansen (1926), 322: "Praeterea, anima bruti non potest appetere [Toivanen (2013a), 315, suggests to read *apprehendere* here] aeternitatem, quia nec potest ipsam cognoscere. Licet enim aliqua futura indendant ac per consequens et praecogitent, puta, pullos habere de ovis vel refectionem ex cibo: non tamen possunt futura praecogitare sub plena ratione futurae aeternitatis, immo nec nisi prout futura sunt quasi praesentialiter correlata et connexa iis quae praesentialiter sentiunt et agunt. Et idem est de recordatione praeteritorum. Non enim potest praeteritio et futuritio plene apprehendi, nisi plene sciantur non esse actu et plene discernantur a praesenti, sicut non ens actu ab ente actu. Bruta autem hoc non possunt, quia nec apprehendunt illa, nisi prout quasi praesentialiter se offerunt in specie et actu imaginationis." On this passage see also Toivanen (2013a), 315f.

CHAPTER 28

Imperfect or Particular Prudence (Albertus Magnus, Thomas Aquinas)

Acting prudently, as has been repeatedly stressed, means to do something in the *present* for the *future* by means of experience from the *past*. Memory and foresight are thus essential cognitive components of prudence. As the previous chapters have shown, nonhuman animals do not fully possess these capacities, according to various later medieval thinkers. Although many animals are capable of *remembering* things, nonhuman species are incapable of what was usually called *recollection*. Hence, their memory is incomplete or imperfect in comparison to human memory, as, for instance, Thomas Aquinas and Albertus Magnus put it. With the component of foresight, it is even worse because all of the authors under consideration argued that other animals do not anticipate future states or events. All they do is to react to present stimuli. Admittedly, their behaviours are often *directed towards* the future. But this does not mean that they actually *discern* between the present and the future. Rather, they provide for the future without really knowing that they will profit from their present operations. In both cases, the lack of intellect and reason is what causes the deficiency, or even lack, of recollection and foresight. The question now is what conclusions should be drawn from this for the prudence of nonhuman animals. Can they be prudent if they do not, or not fully, possess memory and foresight?

One possible answer to this question was given by Albertus Magnus. For Albert, it is clear that memory is a necessary prerequisite of prudence. Unless there is a storage of sensible forms and insensible intentions, there is no prudence, he argues. Consequently, those animals which possess sensory powers but lack memory, are 'imprudent' (*imprudentia*). All other animals can be called 'prudent' (*prudentia*) at least to the extent that they possess 'a control of life from memory' (*regimen vitae ex memoria*).[1] One might assume that Albert says this simply because he is trying to cope with the famous statement at the

1 See Albertus Magnus, *Metaphysica*, lib. I, tr. 1, c. 6, ed. Geyer (1960), 9: "Quibusdam vero ex sensu secundum actum facto fit memoria prius acceptorum sensibilium. Et cum memoria non tantum sit thesaurus et coacervatio formarum sensibilium prius acceptarum, sed etiam intentionum convenientis et inconvenientis, boni et mali, amici et inimici et huiusmodi cum sensibilibus ab aestimativa acceptorum, sicut in libro *De memoria et reminiscentia*

 | DOI 10.1163/9789004363779_035

beginning of the *Metaphysics* in which Aristotle claims that there is memory in some animals, but not in others and thus that this is also the case with prudence.[2] Yet, Albert's account of "animal prudence," as Summers calls this capacity, is much more than just a literal adoption of Aristotle's text because Albert is well aware of the deficiency of both memory and foresight in nonhuman animals.[3] Therefore, he needs to explain how there can be prudence if the lack of intellect and reason affects a creature's cognitive components.

In his *Metaphysics* commentary, Albert solves this problem by saying that we can call nonhuman animals prudent but 'not according to the perfect measure of prudence' (*non secundum perfectam prudentiae rationem*). If we talk about prudence in bees, for instance, we do not talk about 'an active habit involving true reason' (*activus habitus cum ratione vera*) which aims at gathering whatever is useful or harmful for life.[4] In other words, what one finds in them is prudence 'in a certain respect' (*secundum quid*) rather than prudence 'in an absolute sense' (*simpliciter*), as he puts it in his commentary on Aristotle's *Ethics*.[5] In order to illustrate this difference, he refers to the example of the reflection of light or sound. If we look at a ray of sunlight that is reflected by a wall, we still see a ray of sunlight but this ray is less intense than the unreflected ray. The same applies to prudence in human and nonhuman animals: in comparison to humans, nonhuman animals possess a weaker form of prudence. But in what sense is the prudence of nonhuman animals *weaker* than that of humans?

According to Albert, there are two main differences between human and nonhuman prudence.[6] First, there is a difference with regard to the mental representations that are involved in prudent behaviour, or, more precisely, in the cognitive processes underlying such behaviour. While humans employ not only particulars but also universals, other animals do not go beyond the level

---

probavimus: propter hoc quidem alia prudentia sunt animalium habentia regimen vitae ex memoria, alia vero solo sensu vigentia sunt imprudentia."

2 See p. 165 n3.

3 On "animal prudence" see Summers (1987), 270.

4 See Albertus Magnus, *Metaphysica*, lib. I, tr. 1, c. 6, ed. Geyer (1960), 9: "Dico autem memorantia prudentia non secundum perfectam prudentiae rationem, quae est activus habitus cum ratione vera eorum quae in nobis sunt, ad vitam conferentium."

5 See Albertus Magnus, *Super Ethica commentum et quaestiones*, lib. VI, lec. 10, ed. Kübel (1987), 462: "Dicendum quod in brutis non est prudentia simpliciter, sed secundum quid. Unde dicit Commentator, quod est in eis quaedam resonantia prudentiae. Fit autem resonantia ex soni reflexione; unde reflexum procedit in debilius esse sicut lumen defectibilius." See also Roling (2011), 232.

6 On these differences see also Köhler (2014), 140f.

of particulars. Even if they grasp insensible intentions, such as the harmfulness or usefulness of an object, they sense a particular, as has been shown in Part 2. Moreover, they do nothing without immediate sensory input. For instance, sheep do not run away as long as there is no sensory stimulus, such as a wolf, triggering a reaction.[7] Similarly, memory and apparently foresightful behaviour requires certain present stimuli such as an appetite or the perception of a certain change in the environment, as we have seen in the previous chapters. And in this sense nonhuman prudence is weaker than human prudence.

The second difference is the cognitive power (*virtus*) that is involved in this process. Human prudence derives from the intellectual part of the soul, whereas the prudence of nonhuman animals is based on the powers of the sensory part.[8] Therefore, there is no (or less) flexibility and variability in the operations of nonhuman animals. Put differently, prudent actions in nonhuman animals are restricted to a certain set of operations which are given to them by nature, and this set is the same for all members of a species. All ants are prudent in the same way, for example, not only because they all store grain for the future, but also because they do so in much the same way. Similarly, certain aquatic animals, as well as serpents and creeping animals, are prudent in that they ambush prey.[9] They are prudent in the same way as other members of their species because their predatory behavior is carried out in much the same fashion. Humans, of course, also engage in the same prudent behaviors as other members of their species. But they stand in contrast to nonhuman animals in that their prudent actions take different forms. Whereas all octopuses hunt using a small array of techniques, human hunting assumes a far greater range of forms, from the use of sticks or spears to the use of high-powered rifles.

7 See Albertus Magnus, *Super Ethica commentum et quaestiones*, lib. VI, lec. 10, ed. Kübel (1987), 462f.: "Debilitatur autem prudentia in brutis ab eo quod est in nobis, dupliciter: et quantum ad formam moventem et quantum ad virtutem moventem. Quantum ad formam, quia prudentia est in nobis formis simplicibus et universalibus separatis ab individuantibus; unde procedit quodam syllogismo accipiens particulare sub universali. Sed bruta accipiunt formas, quibus diriguntur in opere, quae sunt intentiones nocivi et conferentis, cum materia et non abstrahunt ab individuantibus; unde etiam eorum operatio semper incipit a sensibus, sicut patet, quia ovis non fugit lupum, nisi quando videt eum."

8 See Albertus Magnus, *Super Ethica commentum et quaestiones*, lib. VI, lec. 10, ed. Kübel (1987), 463: "Quantum autem ad virtutem moventem, quia homo operatur secundum partem animae, quae est dominativa sui actus; unde etiam diversi diversimode operantur. Sed principium operationis brutorum est anima, inquantum est natura quaedam, et ideo omnia similia speciei habent easdem operationes." This is also mentioned in *De animalibus*, lib. VIII, tr. 6, c. 1, ed. Stadler (1916), 666.

9 See Albertus Magnus, *De animalibus*, lib. XXI, tr. 1, c. 7, ed. Stadler (1920), 1343.

In short, there is a certain uniformity in the prudent behaviour of nonhuman animals and diversity in the prudent actions of humans. Nonetheless, both human and nonhuman animals behave prudently in one way or another.

Like Albert, Thomas Aquinas also holds that prudence can be found in other animals, and he agrees that their prudence is a weaker form of human prudence or 'imperfect prudence' (*prudentia imperfecta*), as he calls it. It is imperfect mainly because whatever nonhuman animals do is done without the help of intellect or reason.[10] Therefore, only humans are capable of 'deliberating by virtue of reason' (*ex ratione deliberare*) about what should be done for achieving a certain end. All other animals, in contrast, do not go through a deliberative process, but rather make a judgment 'by virtue of some instinct of nature' (*ex quodam naturae instinctu*). They are endowed with the power of 'natural estimation' (*aestimatio naturalis*), and by virtue of this power they judge about what should be fled or sought for the sake of survival.[11] These judgments differ, however, from the judgments of humans, as we have seen in Chapter 15, because we do not only judge by our external and internal senses, but we also form judgments by intellect and reason. In contrast to sensory judgments, intellectual judgments have a predicative structure, and thus they are not instinctual or 'natural judgments' (*iudicia naturalia*). Consequently, our prudence is different from the prudence of nonhuman animals in that we have more than just 'natural prudence' (*prudentia naturalis*).[12]

---

10 See Thomas Aquinas, *Sentencia libri De sensu et sensato*, tr. 2, c. 1, ed. Leonina XLV.2 (1985), 103f.: "Ita etiam in progressu ab animalibus ad homines quedam inveniuntur in quibus aliqua similitudo rationis appareat: cum enim prudencia sit propria virtus hominis (est enim prudencia recta ratio agibilium, ut dicitur in VI Ethicorum), inveniuntur quedam animalia quamdam prudentiam participare non ex eo quod habeant rationem, sed ex eo quod instinctu naturae moventur per apprehensionem sensitivae partis ad quedam opera facienda, ac si ex ratione operarentur."

11 See Thomas Aquinas, *In duodecim libros Metaphysicorum Aristotelis expositio*, lib. I, lec. 1, n. 11, eds. Cathala & Spiazzi (1964), 7f.: "Dicitur autem prudentia aliter in brutis animalibus, et aliter hominibus inesse. In hominibus quidem est prudentia secundum quod ex ratione deliberant quid eos oporteat agere; unde dicitur sexto Ethicorum, quod prudentia est recta ratio agibilium. Iudicium autem de rebus agendis non ex rationis deliberatione, sed ex quodam naturae instinctu, prudentia in aliis animalibus dicitur. Unde prudentia in aliis animalibus est naturalis aestimatio de convenientibus prosequendis, et fugiendis nocivis, sicut agnus sequitur matrem et fugit lupum." For similar passages see *Quaestiones disputatae De veritate*, q. 25, a. 2, corp., ed. Leonina XXII.3.1 (1973), 733; *Sentencia libri De anima*, lib. II, c. 29, ed. Leonina XLV.1 (1984), 190. See also Deeley (1971), 61; Manzanedo (1989), 107; Roling (2011), 236f.

12 See Thomas Aquinas, *Summa theologiae* II-II, q. 45, a. 15, co., ed. Leonina VIII (1895), 363: "Praeterea, prudentia magis convenit naturae humanae quam naturae brutorum

Aquinas also calls such natural prudence 'particular prudence' (*prudentia particularis*) because, like Albert, he thinks that it differs from the 'universal prudence' (*prudentia universalis*) of humans in two important respects. First, the prudent operations of nonhuman animals are based on the cognition of particulars, whereas the prudent actions of humans might also (and most often do) include or even lead to the cognition of universals. Second, human prudence is universal insofar as it is not limited to particular operations. We can make prudent decisions when providing for future meals, for instance, but we can also be prudent in numerous other cases; that is, we can use our prudence quite flexibly or universally. This capacity is not found in other animals. Their prudence is restricted to those actions which are given to them by nature, such as the ants' provision for the winter.[13] For this reason, such prudence is *natural* and *particular*.

With respect to the differences between human and nonhuman prudence, one might ask whether it makes sense to call nonhuman animals prudent at all. Why should we take their behaviour to be prudent if the cognitive processes underlying this behaviour significantly differ from those processes underlying our prudent actions? Aquinas shows some awareness of this problem when he writes that there is only a 'likeness of prudence' (*similitudo prudentiae*) in nonhuman animals.[14] In other words, ants and other nonhuman animals

animalium. Sed bruta animalia habent quasdam naturales prudentias; ut patet per Philosophum, in VIII *de Historiis Animal.* Ergo prudentia est naturalis. [...] Ad tertium dicendum quod in brutis animalibus sunt determinatae viae perveniendi ad finem: unde videmus quod omnia animalia eiusdem speciei similiter operantur." See also *Quaestiones disputatae De veritate*, q. 15, a. 1, ed. Leonina XXII.2 (1972), 480.

13 See Thomas Aquinas, *Sentencia libri De sensu et sensato*, tr. 1, c. 1, ed. Leonina XLV.2 (1985), 13f.: "Circa primum considerandum est quod prudencia est directiva in agendis, et universalis quidem prudencia est directiva respectu quorumcunque agendorum, unde non est in aliis animalibus nisi in solis hominibus, qui habent rationem universalium cognoscitivam, et in aliis autem animalibus sunt quedam prudencie particulares ad aliquos determinatos actus; sicut formica que congregat in estate cibum de quo vivat in hyeme. Predicti autem sensus, maxime auditus et visus, proficiunt animalibus ad huiusmodi prudencias particulares et hominibus ad prudenciam universalem ad hoc quod aliquid bene fiat [...]." See also Hünemörder (1988), 201; Di Martino (2008), 99.

14 See Thomas Aquinas, *Sentencia libri De sensu et sensato*, tr. 2, c. 1, ed. Leonina XLV.2 (1985), 104: "Pertinet autem ad prudenciam, ut prudens dirigatur per eam in his quae imminent sibi agenda ex consideratione non solum presencium, set etiam preteritorum [...]. Unde et in aliis animalibus in quibus invenitur prudencie similitudo participata, necesse est esse non solum sensum presencium, sed etiam memoriam preteritorum. Et ideo philosophus dicit in principio Methaphisice quod quibusdam animalibus ex sensu memoria fit, et propter hoc prudencia sunt."

are prudent only insofar as they are capable of remembering the past and insofar as they have 'a certain anticipatory power with regard to their own life' (*quandam potentiam provisivam circa propriam vitam*). But since their prudence does not originate from rational faculties, it is just prudence 'by a certain likeness' (*per quandam similitudinem*).[15] Still, he would not go as far as saying that other animals are totally devoid of prudence. Rather, he points to the structural differences that exist between the prudent behaviours of different animal species. In this sense, both Albert and Aquinas acknowledge that, despite all differences, at least some nonhuman animals are prudent, as well.

15 Thomas Aquinas, *Sententia libri Ethicorum*, lib. VI, lec. 6, ed. Leonina XLVII.2 (1969), 353: "Ille enim qui potest bene speculari singula quae pertinent ad se ipsum dicitur esse prudens et tali conceditur sivi attribuitur prudentia. Et inde est quod per quandam similitudinem homines dicunt quasdam bestias esse prudentes, quaecumque scilicet videntur habere quandam potentiam provisivam circa propriam vitam, non quidem ex ratione, quod proprie ad prudentiam pertinet." See also Manzanedo (1989), 106.

CHAPTER 29

Prudence by Analogy (Giles of Rome, John Duns Scotus)

Albertus Magnus and Thomas Aquinas have answered the question of whether other animals are prudent in the following way: they are prudent, but there are crucial differences between human and nonhuman prudence, most of all that our prudence derives from the operations of intellect and reason. The prudent behaviour of nonhuman animals, by contrast, is based on the operations of sensory powers and, in many cases, it is their instincts which make them behave prudently. Thus, they do not recollect past experiences and on the basis of this recollection deliberate about what should be done in the present in order to profit in the future. While we remember the last time we went to the grocery store and found it closed and, therefore, conclude that we should go to the supermarket earlier next time, the ant collects and stores grains simply because this is how it has been 'programmed' by the divine creator, so to speak. Nonetheless, ants can be called prudent because they provide for the future and so will profit *later* from what they do *now*.

The argument that nonhuman animals can be called prudent even though their prudence differs from human prudence in various respects, was heavily criticised by other authors, especially Giles of Rome and John Duns Scotus. Both of them explicitly discuss the question of prudence in their commentaries on Aristotle's *Metaphysics*. Giles immediately rejects the view that there is not 'prudence properly speaking' (*prudentia proprie dicta*) in nonhuman animals, but instead some substitute for prudence, namely 'some natural sagacity or natural instinct' (*quedam sagacitas naturalis sive quidam instinctus naturalis*), by virtue of which they engage in prudent behaviour.[1] This is a rough sketch rather than a minute account of Albert's and Aquinas' position but both

1 Giles of Rome, *Questiones metaphisicales* (Venice 1501), fol. 8ra: "Ad istam questionem respondent breviter aliqui et dicunt quod prudentia inventa in brutis non est prudentia proprie dicta: sed prudentia que invenitur in brutis nihil aliud est quam quedam sagacitas naturalis sive quidam instinctus naturalis per quem agunt bruta ea que agunt. Et hoc declarant per exemplum: sicut formica existens in estate habens memoriam famis preterite in hieme congregat fructum et ea ex quibus possit vivere in hieme: et similiter hirundo facit nidum suum quadam sagacitate naturali." Note that Giles' commentary has not yet been edited. I, therefore, rely on my own transcription of the 1501 print from Venice. Large parts of this question have also been transcribed by Köhler (2014), 146–148.

 | DOI 10.1163/9789004363779_036

of them do indeed claim that in nonhuman animals natural instinct takes over the work which is done by intellect and reason in human beings. For Giles, it is absurd to claim that prudence is nothing but natural instinct because then we are obliged to attribute prudence to *all* animals since all of them are obviously endowed with natural instinct: there is no animal which does not instinctually flee an unpleasant object, for instance, and so all animals react instinctually in one way or another.[2]

Albert and Aquinas could, of course, reply that this is not a proper description of their view. It takes more than simple avoidance behaviour for prudence, because prudent behaviour involves memory and is at least *directed* at the future. But Giles (as well as Scotus) rejects the claim that ants, for example, use their memory when collecting grain during the summer for the coming winter. If the ants' provision were really based on a memory of the past winter, one could hardly explain why even those ants which are born during the summer and thus have not yet experienced the cold season still collect and store grain for the future. Apparently, this cannot be accounted for by referring to memory of the past.[3]

It is difficult to tell how Albert and Aquinas would reply to this argument. They might stress two points here. First, they might stress that even if ants do not use memory in this particular case, they usually use memory 'in place of reason' (*loco rationis*) because, like bees and many other animals, they have fixed dwellings to which they return and where they store their food.[4] Second, the role of memory and foresight must not be overemphasised in the case of

2 See ibid.: "Sine dubio istud quod isti dicunt non habet veritatem. Quoniam si prudentia in brutis esset sagacitas nature cum sagacitas et instinctus nature insint omnibus animalibus brutis quecumque sint illa: tunc omnia bruta animalia essent prudentia: et hoc non est verum. Quod enim omnibus brutis insit sagacitas naturale patet: quoniam non est aliquid brutum animal in mundo quod si pungatur quod non retrahat se."

3 See Giles of Rome, *Questiones metaphisicales* (Venice 1501), fol. 8ra: "Et quod ipsi dicunt quod formica memor famis in hieme congregat in estate falsum est. Nam si modo nasceretur formica et esset estas: modo congregaret non propter famem preteritam sed propter naturalem instinctum."; John Duns Scotus, *Quaestiones super libros Metaphysicorum Aristotelis I–V*, lib. 1, q. 3, §5, eds. Andrews et al. (1997), 87f. The same point is also made by Adam Wodeham, *Lectura secunda in primum librum Sententiarum*, prol., q. 4, §8, ad 1, ed. Wood (1990), 99f. On Giles and Scotus see also Roling (2011), 239f.; Köhler (2014), 146–151.

4 See Albertus Magnus, *Metaphysica*, lib. 1, tr. 1, c. 6, ed. Geyer (1960), 9: "Sed non habentia rationem veram memoria utuntur loco rationis et ordinant aliquo modo suae vitae commodum per quandam civilitatis et felicitatis similitudinem, sicut est videre in apibus et gruibus et multis huiusmodi animalibus; sed tam apes quam grues vigent solum memoria. Cuius signum est, quod a longiquis locis, ad quae transferuntur, revertuntur ad proprias habitationes et casas."

nonhuman prudence, because what really matters is that animals manage to do things in the present from which they will profit in the future, and this is what prudence is about.

Giles admits that ants use memory in collecting and storing grain because, unlike many other animals, they act 'naturally' (*naturaliter*) but not 'uniformly' (*uniformiter*). If an ant finds a good place for collecting grain it will, by virtue of its memory, return to this place as long as some grain is left there. Another ant, however, will not necessarily walk to the very same place. Hence, there is no uniformity in this regard mainly because every ant relies on its own memory.[5] Scotus also agrees with this.[6] Still, he and Giles disagree with authors like Albert and Aquinas on whether it makes sense to attribute prudence to animals like ants. Unlike Albert and Aquinas, they think that a behaviour cannot qualify as *prudent* behaviour if it is based on something like imperfect memory and quasi-foresight. Hence, it does not make sense to speak of something like 'imperfect prudence', in their view. They do, of course, admit that we often call other animals prudent, especially when we are astonished by what they do or how they do something. But we then employ the term 'prudence' only 'metaphorically' (*metaphorice*) or 'by way of analogy' (*similitudinarie*).[7] To be clear,

---

5 Giles of Rome, *Questiones metaphisicales* (Venice 1501), fol. 8rb: "Quedam enim sunt in eis in que bruta feruntur naturaliter et semper uniformiter sicut dicit philosophus in 2° phisicorum quod aranea facit telam naturaliter et semper uniformiter: et hirundo nidum: [...] aliqua vero sunt in brutis in que bruta non feruntur naturaliter nec semper uniformiter sed in talibus convenit error et difformitas: sicut patet quod una formica congregat in illo loco et alia formica in alio loco. Non enim omnes formice congregant in eodem loco et in talibus in quibus bruta non agunt uniformiter egent aliqua regula que dirigat ipsa bruta in tales operationes et illa regula est memoria. Videmus enim quod una formica congregat unum granum frumenti in uno loco et alia granum frumenti in eodem loco et non in alio et hoc est quia illa formica habet memoriam de loco in quo posuit primum granum. Videmus etiam quod formica vadit iterum ad acervum a quo accipit primum granus. Et istud est quia habet memoriam quod inde accepit primum granum. Unde ipsa memoria est prudentia in brutis."

6 See John Duns Scotus, *Quaestiones super libros Metaphysicorum Aristotelis I–V*, lib. I, q. 3, §6, eds. Andrews et al. (1997), 88: "Prudentia autem sicut in nobis est habitus consiliativus, non de fine sed de his quae sunt ad finem, non circa necessaria sed contingentia, ita et in illis est circa illa quae possunt sic et aliter facere. Puta quod congreget vel reponat in hoc loco vel illo, et ex hoc cumulo vel illo – ex memoria loci ubi primum granum reposuit, et cumuli unde primum sustulit –, et quod aranea magis facit telam in loco ubi maior copia muscarum, vel hirundo facit nidum ubi est difficilior accessus." Note that Scotus, unlike Giles, does not think that spiders and swallows build their webs and their nests uniformly.

7 John Duns Scotus, *Quaestiones super libros Metaphysicorum Aristotelis I–V*, lib. I, q. 3, §5, eds. Andrews et al. (1997), 87, and §14, 91; Giles of Rome, *Questiones metaphisicales* (Venice 1501), fol. 8ra.

Giles and Scotus do not argue that one should totally refrain from calling other animals prudent. But, in their view, this is just a way of speaking, which has little, if not nothing, to do with the actual cognitive capacities that underlie our own prudent actions. In this respect, one could say that their concept of prudence is more demanding than the notion employed by Albert and Aquinas because while they think that prudence necessarily requires the powers of intellect and reason, Albert and Aquinas say that prudence can, to some extent, be found in those animals that lack those faculties, and so, in their view, there is prudence on both sides of the animal/human boundary.

However, one should not make the mistake of claiming that one account (e.g. the theory of Albert and Aquinas) is clearly superior to another account (e.g. the explanation of Scotus and Giles). Rather, one should pay attention to the fact that they all start from the same premise: all of them deny the faculties of intellect and reason to nonhuman animals, and so they agree that in this sense only humans are *rational animals*. However, they disagree on the question of what this denial of intellect and reason amounts to. While Scotus and Giles argue that this establishes a clear line between us and nonhuman animals, Albert and Aquinas take this line to be less clear as far as prudence is concerned. Thus one could say that they choose a different explanatory strategy. As we have seen in Parts 2 to 4, similar explanatory differences can also be found with regard to other rational capacities. We shall now take a closer look at them in the subsequent and final part of this study.

PART 6

Rationality without Reason?

Introduction to Part 6

The previous parts have examined various later medieval theories of animal rationality. Such theories, it has been said, mainly arose from the grey areas of the Aristotelian model of the soul and its implicit definition of the animal/human boundary. According to this definition, humans are rational animals. They are rational, first and foremost, because they are endowed with rational *faculties*, namely, intellect and reason. Second, they are rational because these faculties enable them to engage in rational or intellectual *processes* such as concept formation, judging, and reasoning. It is this triad of intellectual operations that shapes the way in which humans perceive and interact with the world. In many cases, humans behave rationally because their behaviour is based on a rational cognitive process. For instance, before taking our dog out for a walk we have reasoned that it is a good thing to do so because otherwise it might ruin the carpet. Theories of animal rationality, however, concern the question of whether this is something only we can do. Why should one assume that our dog cannot do this? Why should one assume that it is completely non-rational? If it is non-rational how can one explain that it gets excited about all sticks, and so seems to generalise? Or that it falls prey to sensory illusions, and so seems to judge? Or that it takes a detour in order to get around an obstacle, and so seems to reason?

It is questions like these to which theories of animal rationality try to provide an answer, and since late medieval thinkers obviously dealt with such questions it makes sense to say that they developed such theories. What remains to be seen, however, is how their theories compare to the theories of our own age. The commonly accepted narrative has it that later medieval theories differ radically from contemporary theories of animal rationality. Many would even say they differ to such a degree that there is no medieval analogue or predecessor of contemporary theories of this kind at all. Whether or not this is the case will be the subject of this last part. It will begin by sketching the differences that exist (Chapter 30), among the most important of which surely is that later medieval theories were developed within the framework of Aristotelian psychology. Although Aristotle's model of the soul is no longer seriously defended by anyone in contemporary animal psychology, there is a certain continuity between later medieval and contemporary theories in the sense that they usually deal with the question of how to describe the cognitive processes that underlie a certain behaviour (Chapter 31). Apparently, this can be done in many different ways which is why there exist not just one, but many theories of animal rationality. In other words, there are different answers to the

 | DOI 10.1163/9789004363779_037

question of whether those cognitive processes are radically different or quite similar in human and nonhuman animals. Those who claim that they are radically different are called 'differentialists' in contemporary animal philosophy. 'Assimilationists', by contrast, are those who focus on the similarities rather than the differences. To be clear, assimilationists and differentialists agree that there are both differences *and* similarities. Still, they disagree on whether the former outweigh the latter or vice versa.

This disagreement is something one can also find among later medieval authors (Chapter 32). As various examples show, they adopted different strategies when trying to account for the cognitive processes that underly a certain behaviour. While some considered those processes to be radically different, others stressed the similarities that exist between human and nonhuman animal cognition. This in itself is not much of a surprise because there is hardly any topic on which later medieval thinkers spoke with one voice. What is surprising is that some of them seriously considered the option of what could be called 'rationality without reason' (Chapter 33). This means that they in some sense dissolved the link between faculty and process rationality that usually characterises the medieval concept of rationality. For instance, they claimed that a process, such as reasoning, that was usually taken to require a rational faculty, can also be engaged in without this faculty.

There are, of course, important differences between rationality with and without reason but what is interesting to see is that the latter very much resembles the kind of practical and domain-specific rationality that is attributed to nonhuman animals in the contemporary debate. Besides the richness of the theoretical landscape in the later Middle Ages that is often ignored, this particular theoretical option establishes an interesting connection between later medieval and contemporary theories of animal rationality. Yet, before we can begin to closely examine this connection and the parallels between the theories of different periods it is necessary to briefly look at the differences that exist between thirteenth- and fourteenth-centuries theories and contemporary theories of animal rationality.

CHAPTER 30

Medieval and Contemporary Theories: The Differences

Later medieval and contemporary theories of animal rationality differ in various respects. One of the most obvious differences certainly is that they were developed in different contexts. By and large, these contexts are academic ones. However, one must not ignore that life in medieval academia was not exactly the same kind of life many people use to live in modern universities. With regard to natural philosophy and to what we nowadays call psychology, it is important to note that large parts of the academic discourse on animals in medieval universities arose from the Aristotelian commentary tradition (see Chapter 4). There is hardly any scholar in the later Middle Ages who did not have contact with Aristotelian philosophy in one way or another. This is not to say that commentaries on Aristotle's writings were the only place where this discourse took place. As many of the examples mentioned in the previous parts have shown, questions concerning the differences between human and nonhuman animal cognition also came up in theological treatises or other writings which were not primarily concerned with any of Aristotle's ideas. Nevertheless, those ideas influenced later medieval authors' views, and so one can come across elements of Aristotelian psychology also in works on, say, Trinitarian theology. The modern discourse on animals and their cognitive capacities also spans various disciplines, such as philosophy, psychology, biology, and ethology. Yet, it has not been as influenced by something like the Aristotelian commentary tradition as was the later medieval discussion.

A second difference is that this discourse has now developed into a subject in its own right. There is, of course, not a single discipline that is solely concerned with animal cognition these days, but the latter has now become a stand-alone topic of scientific and philosophical enquiry like many other topics in contemporary scholarship. In the later Middle Ages, the situation was different insofar as the discussion on nonhuman animals and their cognitive capacities did usually not begin as such. Take the example of Adam Wodeham's and Gregory of Rimini's debate over animals' capacity of judging (Chapter 18). It is very illuminating in this context, as it is carried out in two theological treatises, namely, their commentaries on Peter Lombard's *Sentences*. It actually began as a debate over the kind of knowledge of God humans can have but then gradually developed into a debate over the nature and origin of sensory illusions. In this

 | DOI 10.1163/9789004363779_038

context, Adam Wodeham introduces the dog aboard the ship as an argument against the position that sensory illusions originate from erroneous intellectual judgments (which is actually the view he wants to defend). The dog is thus used as an *explanans* for the sensory nature of illusions because the argument runs as follows: (P_1) dogs lack intellect and reason; (P_2) dogs fall prey to optical illusions; therefore, (C) optical illusions have sensory rather than rational or intellectual origins. However, in the course of the discussion, the dog's behaviour and its cognition are successively turned into the *explanandum* and give rise to a debate over whether nonhuman animals are capable of judging and should be called rational.

A similar case is Roger Bacon's account of animal cognition (chapters 11 and 20). The context is not theological but physical or, more precisely, optical and the question in this case is whether certain modes of vision originate from the sensory or the rational part of the soul. Like Adam Wodeham, Roger Bacon introduces nonhuman animals as an *explanans* in the first place. What he wants to show is that these modes of vision (which, in the Latin translation of Alhacen's *De aspectibus*, are called 'by sense alone', 'by knowledge', and 'by syllogism') do not require rational powers. The argument runs like this: (P_1) the three modes of vision can be found in many nonhuman animal species; (P_2) nonhuman animals lack the rational part of the soul; therefore, (C) the three modes of vision originate from the sensory soul. Although this is a very straightforward argument, Bacon has to explain how nonhuman animals can actually engage in more complex cognition, such as 'vision by knowledge' or 'by syllogism', and so his argument turns into an examination of those processes on which such cognition is based in nonhuman animals. Nonhuman animals and their cognition now become the *explanandum*. As has been shown, Bacon suggests that there is a cogitative power in such animals which is something that was usually denied to them because it was commonly considered to be a rational faculty. So a discussion about optics turned into a discussion about animal rationality.

In both cases, there is an interesting transformation going on. Nonhuman animals are introduced because they are supposed to prove the *sensory* character of a particular cognitive phenomenon. One could say, they function as a sort of litmus test (see Chapter 4). The crucial point now is that this would not work without a very strong anthropological presupposition, namely, the premise that nonhuman animals lack rational faculties. The concept of *faculty rationality* (which is based on the idea of Aristotelian faculty psychology) obviously plays a fundamental role here. This establishes a third difference between later medieval and contemporary theories of animal rationality.

Still, in the course of this study it has become clear that this concept came under attack. The more nonhuman animals were used as litmus tests on different occasions, the more it was debated whether those tests actually proved what they were supposed to prove. It did not simply suffice to stipulate that nonhuman animals lack rational powers. Rather, it became necessary to explain how they can actually engage in various kinds of cognition, both simple and more complex, if they do not have these powers. And so they gave rise to a debate over the link between *behaviours*, *faculties*, and *processes* within Aristotelian faculty psychology (see Chapter 7). The general question was, first, whether one and the same behaviour is based on one and the same cognitive process and, second, whether this process always requires one and the same faculty. Although it might seem that this question is only relevant within this particular theoretical framework, it is actually something that brings us to the commonalities between later medieval and contemporary theories of animal rationality because both deal with very similar (if not the same) kinds of problems in the end.

CHAPTER 31

Medieval and Contemporary Theories: The Commonalities

One of the main problems both medieval and contemporary theories deal with is the question of what kind of *process* underlies a certain *behaviour*. We observe the behaviour of something – a machine, a plant, a human, or nonhuman animal – and wonder what it is that brings about this behaviour. For instance, plants and animals grow. Yet, is it the same kind of process that makes all living beings grow? Or, to give another example, robots and humans can raise their arms, but is what is going on in the robot when it moves its arm the same kind of process that is going on in humans when we raise our arm? Another example (which has already been mentioned in Chapter 7) would be this: we confront a chimpanzee and a human child with a transparent tube at the bottom of which lies a peanut. As the tube is quite long and narrow, neither the chimpanzee nor the child manage to grab the peanut with their fingers. Yet, after a while both begin to fill water into the tube until the rising water level brings the peanut within reach. So both the chimpanzee and the child manage to find the right tool for getting the peanut.[1] Now does this mean that the cognitive processes guiding their behaviour are basically the same? One could go on endlessly with this list of examples and questions but in each and every case the fundamental question would be the same: What kind of process underlies the behaviour?

Quite often, we are inclined to think that similar behaviours are based on similar processes. As said before, this inclination is expressed by what is called the 'argument from analogy' (see Chapter 7). This argument suggests that there is an analogy between the behaviours we observe and the processes underlying those behaviours. However, this argument gives rise to various – and sometimes rather serious – questions. One of these questions, with a more ethical impetus in this case, would be this: if a robot manages to raise its arm like we do and if a chimpanzee succeeds in finding the right tool like a human infant does why don't we grant the robot and the ape the same kind of rights we grant to our fellow humans? Surely, this question only arises when we assume that there is a link between cognitive capacities and moral status but usually we

1 On this so-called 'floating peanut task' see Hanus et al. (2011).

 | DOI 10.1163/9789004363779_039

make this assumption and so did later medieval thinkers.[2] And even if one does not care about the argument's ethical implications another question remains: are we just as smart as robots or chimpanzees? Is it really the same kind of process that makes us and robots raise our arms or that makes us and chimpanzees find the right tool?

In order to find an answer to these questions both medieval and contemporary scholars try to look at the processes underlying behaviours as closely as they can. Now one might think that contemporary scholars have a serious advantage in this regard for at least three reasons. First, they have the help of evolutionary biology and genetics. Both strengthen a principle that supports the argument from analogy, namely, the so-called 'principle of continuity'. According to this principle, there is more psychological continuity between different beings the more they are related to each other from an evolutionary perspective.[3] This relation can be determined by studying the physiology of different beings but genetics provide a fundament that is even stronger. Consequently, similar behaviours in closely related species, say humans and great apes, are likely to originate from similar cognitive processes because the latter take place in very similar physiological structures. Of course, there is no guarantee that similarity in structure always leads to similarity in cognition. Still, the evolutionary perspective adds a new dimension to the study of similarities in behaviour that medieval theories lack.

Second, contemporary theories seem to have the upper hand as they can rely on a huge body of empirical findings. Comparative psychology and ethology are relatively young disciplines, but within decades they have multiplied the number of systematic studies and refined the methods with which behaviour is studied.[4] Medieval thinkers, in contrast, mainly drew on anecdotal material. In many cases, they did not even make observations themselves but simply reproduced examples of animal behaviours they found in the literature.[5] The

2 On this connection in contemporary philosophy see the contributions to Petrus & Wild (2013), esp. Glock (2013). An overview of animal ethics in the Middle Ages provides Oelze (forthcoming a). The connection might seem to be less evident with regard to robots or machines but the intensified research in things like autonomous cars has given rise to a controversial debate over the moral status of such artificial beings; see, for instance, Wallach & Allen (2008). Interestingly, a forerunner of this debate can already be found in the early modern period; see Cheung (2010).

3 See, for instance, De Waal (1999), esp. 258f.; Tomasello (2014), 15. For a critical evaluation see Sober (2000) and (2005).

4 For a short history of this development see Wynne (2007).

5 On the role of observation in comparison to literary knowledge in the later medieval debate over animal cognition see Köhler (2008), 267–272.

nineteenth-century pioneers of evolutionary animal psychology and ethology, Charles Darwin and George Romanes, did of course rely on anecdotes, as well. But during the twentieth century their comparatively lax standards have been replaced by those of international and interdisciplinary empirical research which is hardly comparable to nineteenth-, let alone thirteenth- and fourteenth-centuries studies on animal cognition.

Third, modern research in animal cognition is no longer obliged to rely on the observation of behaviour alone. Instead, it can use the latest techniques of brain imaging. Modern researchers, therefore, seem to have opened a 'window to the mind'. In contrast to later medieval scholars, they can look into the brains of humans and nonhuman animals, and so they can literally see what is going on when an infant and a chimpanzee choose a tool. Generally speaking, it seems as if they have thus found a solution to what might be called the *problem of the inaccessibility of mental states*, also known as the 'inobservability thesis'.[6] For later medieval philosophers this was a serious problem. In the debate over nonhuman animals' capacity of judging, Adam Wodeham nicely captures this difficulty, saying that 'we cannot know if they possess this [capacity], unless by guessing from the effects and movements that follow such simple visions'.[7] In other words, Wodeham cannot look into an animal's head and see its mental states or cognitive processes. All he can do is to observe the effects of those processes, that is, the animal's behaviour.[8]

However, it is rather naïve to think that this problem has been overcome by the methods of evolutionary biology, empirical studies in comparative psychology, and the techniques of brain imaging in modern neuroscience. Admittedly, genetic analyses, systematic empirical studies, and brain imaging techniques are a major advance in the field of animal psychology. But they still do not solve the problem of the inaccessibility of mental states because empirical findings and the images of brain processes need to be interpreted in one way or another. When we empirically test whether an animal is capable of solving a problem, we still cannot directly access the cognitive process that allows it to solve this problem. We do not see the thoughts or beliefs the chimpanzee has when picking a tool. And even if we see that there is a particular neural activity

6 For a discussion see Overgaard (2015).

7 Adam Wodeham, *Lectura secunda in primum librum Sententiarum*, prol., q. 4, §8, ed. Wood (1990), 99: "[...] utrum eis conveniat, scire non possumus, nisi coniciendo ex effectibus et motibus sequentibus tales visiones simplices."

8 This problem was also acknowledged by early-modern thinkers, such as Descartes, who conceded that one cannot prove the absence of thoughts in nonhuman animals since one cannot look into their hearts; see Wild (2006), 189.

in a certain region of the chimpanzee's brain while it picks the tool, we must still determine whether what is going on in this area of its brain is a process such as reasoning or something else. Thus, "we continue to struggle with the question of how to judge the mental life of other species," as Randolf Menzel and Julia Fischer, who both work on contemporary animal psychology, admit.[9]

Now one might object that this problem is not peculiar to theories of cognition but is a problem that any kind of scientific theory has always had and still has to face. For example, a chemist who tries to explain why salt dissolves in water whereas oil does not (or at least not as easily as salt) also has no direct access to every dynamic that is or might be involved in this chemical process. She might, of course, argue that it makes sense to postulate the existence of molecules because the difference in the molecular structures of salt and oil helps to explain why they react differently. She might even show us these molecules under a microscope. Still, she cannot be entirely sure that they cause these differences. Hence, it seems right but trivial to say that both medieval and contemporary theories of animal rationality need to deal with the inaccessibility of what they try to explain because any other theory – no matter whether it concerns mental states, molecular structures, or anything else – in some sense faces this problem.

On the one hand, this is indeed a legitimate objection, because it belongs to the nature of a theory to make certain claims about how things are without actually knowing whether this is the correct explanation. On the other hand, theories of cognition in general and of rationality in particular, belong to a peculiar genre of theories. Their peculiarity can be illustrated by Daniel Dennett's distinction between three different stances in the explanation of behaviour. The first stance is what Dennett calls "the *design stance*."[10] A design-stance prediction of a computer's behaviour is, for instance, when we say something like 'If you press the button which has a B on it, a B is going to appear on the screen of your computer'. The reason is that the computer has been designed to show a B on the screen when the B button is pressed. The second stance is the so-called "*physical stance*."[11] A physical-stance explanation is when we say something like 'If you press the B button but your screen remains black this might be because your computer is not plugged in'. In this case, a physical explanation is given for why the computer does not function as it has been designed. Both kinds of explanations can also be applied to the behaviours of animals. We could, for instance, say that an animal moves from one corner of

9 Menzel & Fischer (2011), 1. See also Wasserman & Zentall (2006), esp. 6f.

10 Dennett (1971), 87f.

11 Ibid., 88f.

the room to another because blood is transported from the heart to its feet and because the muscles in its legs go through a serious of successive contractions and extensions.

The question, however, is whether this completely accounts for the animal's movement. As regards the design stance and the physical stance, the explanation is complete. But is the explanation of *why* the animal begins to move complete? What if the animal begins to move its legs because it *sees* the feeding dish standing in the other corner, *desires* to empty it, and *believes* that this will satisfy its desire? In this case, we do not only refer to the animal's design and the laws of physics but we give an explanation from what Dennett calls "the *intentional stance.*" We predict or explain a system's behaviour "by ascribing to the system the possession of certain information and by supposing it to be directed by certain goals."[12] Usually, this is exactly the kind of explanation cognitive scientists and philosophers give. To be clear, they do not claim that physical or physiological explanations are wrong. Rather, they say that those explanations are only one piece of the puzzle. In their view, it often takes more to move from one corner into another than muscle contraction or blood flow. Put in terms Fred Dretske employs, such an explanation requires not only *causes* but *reasons*, too.[13] Such reasons or beliefs or desires are, however, more difficult to detect than the movements of molecules, the flow of blood, or the contraction of muscles. Therefore, the problem of inaccessibility is particularly relevant for theories which refer to mental states and this is something that has not changed from the medieval period to the modern one.

What has not changed either is that from this problem arises a variety of explanations or theories of cognitive processes. There is not only one way to describe the nature of those processes but many. In the current literature, some attempts have been made to depict the differences between those ways and one influential suggestion has been to distinguish between two explanatory strategies, namely, *differentialism* and *assimilationism*.[14] This distinction goes back to Robert Brandom who, in his book *Articulating Reasons*, has argued that two ways can be adopted if one wants to explore how "discursive and non-discursive creatures" compare to each other.[15] One way stresses the continuities between those creatures. The other road concentrates upon the discontinuities. For instance, if one studies the use of concepts in various creatures, one can either "*assimilate* conceptually structured activity to the

12 Ibid., 90.

13 See Dretske (1988).

14 See Perler & Wild (2005), 73; Wild (2013), 36f., and Glock (2016a), 60.

15 Brandom (2000), 2.

nonconceptual activity out of which it arises." Or one can put weight on "what is *distinctive* of or exceptional about the conceptual."[16]

Now one might think that this distinction also works perfectly for later medieval theories.[17] However, there is a serious problem when applying it to these theories because most scholars who draw on Brandom's distinction take it to work as follows: differentialists, they say, consider the difference between humans and other animals to be *qualitative* or *categorical*. To put it in Darwinian terms, the difference is a *difference in kind*, according to this view. Assimilationists, by contrast, consider it to be *quantitative* or *gradual*. In other words, the extent to which humans differ from apes, dogs, and other nonhuman animals is a *matter of degree*.[18] This gives rise to the question of what it means to say that a difference is qualitative rather than quantitative, categorical rather than gradual. The answer that is usually given is that differentialists posit an *anthropological difference*.[19] This means that, for a differentialist there is at least one fundamental feature that belongs exclusively to humans. For instance, she might hold that the faculty of speech is what sets apart humans from all other animals, because, in her view, only a human being has the capacity to form concepts and propositions. Nonhuman animals might communicate, but they do not speak. Another popular candidate for an anthropological difference is reason, and it is this candidate which seems to have dominated over the centuries, especially in the medieval period. This seems to lead to the following conclusion: *all medieval authors are differentialists* because by denying intellect and reason they establish a clear gulf between humans and other species.

To this classification, one might object that later medieval thinkers were well aware of the animal nature of human beings and so emphasised the continuities that exist between all animals. On the ladder of nature, humans stand much closer to dogs or apes than to plants because they share with the former various *cognitive* powers, whereas with plants they only share the vegetative functions of the soul. And so one might, like Pieter De Leemans and Matthew Klemm, think that "the distinction between human and animal behaviour is presented as a question of more or less, rather than a radical difference in

16 Ibid., 3 [italics in the original].

17 Some attempts to apply it to medieval theories have been made by Perler (2006), 96, and Davids (2017), 30f. and 114–116. See also Wild (2006), 1–11, who employs this distinction for evaluating various early modern theories, and Perler (2009) who applies it to Leibniz's theory.

18 See Perler (2009), 75; Davids (2017), 30; Glock (2013), 120, and (2016a), 60. See also Perler & Wild (2005), 73.

19 See Wild (2013), 36.

kind."[20] This view also seems to be supported by Aristotle who, in a passage of the *Historia animalium*, states that "[n]ature proceeds from the inanimate to the animals by such small steps that, because of the continuity, we fail to see to which side the boundary and the middle between them belongs."[21]

However, it was also Aristotle who prominently established a discontinuity between human and nonhuman animals by denying intellect and reason to the latter. Hence, humans are animals, but they are the only animals that have "the 'special somethin'," as Jonathan Bennett called it.[22] And because of this special something 'humans differ from those that are called brute animals more than by the species', as Albertus Magnus neatly expresses it.[23] For instance, the difference between a lion and a horse is smaller than the difference between a human and a horse or a human and a lion, because lions and horses are simply different kinds of *nonhuman* animals. Humans, by contrast, constitute a very peculiar species of animals which, possessing intellect and reason, has something that no other animal species has.[24] Because medieval thinkers find a clear *metaphysical* difference between human and nonhuman animals, they all seem to have chosen the strategy of differentialism instead of assimilationism.[25]

The question, however, is what such a metaphysical differentialism amounts to and whether choosing this strategy necessarily prohibits one from taking the road of assimilationism. Do differentialists necessarily deny any kind of cognitive continuity between human and nonhuman species by choosing differentialism? Do they always claim that there is a deep gulf between us and other animals with a view to cognitive capacities? For Robert Brandom this is not necessarily the case. At least in *Articulating Reasons*, he says that those who take the path of differentialism "*start* by describing a common genus and go on to elaborate differentiae (whether qualitative or in terms of some

20 De Leemans & Klemm (2007), 158.

21 Aristotle, *Historia animalium* VII.1, 588b5-7, ed. and tr. Balme (1991), 61–63. For a concise summary of this idea see Rheins (2015), 390f.

22 Bennett (1989), 4.

23 Albertus Magnus, *De animalibus*, lib. XX, tr. 2, c. 6, ed. Stadler (1920), 1320: "Homo igitur plus quam specie differt ab hiis quae dicuntur bruta animalia. Specie enim differunt leo et equus: plus autem quam specie differt homo ab utroque ipsorum." See also Köhler (2008), 342.

24 König-Pralong (2011), 74, therefore says that for Albert "l'homme est non seulement *au sommet*, il est aussi *au-dessus* de la nature" [italics in the original].

25 As Perler (2014b), 232, claims this is what unites Aristotelians and Cartesians because they both hold that there is a *categorical* difference between human and nonhuman animals.

quantitative ordering by a particular kind of complexity)."[26] The differences they see are not necessarily *qualitative* or *categorical*. The point is rather that, in contrast to assimilationists, they emphasise the differences more than they stress the similarities that exist. Assimilationists, in turn, do not necessarily reject the existence of an anthropological difference, as Markus Wild points out. Assimilationists might claim that there is much continuity on the *scala naturae* but, at some point, 'the one decisive step' occurs.[27]

This point is crucial for the evaluation of later medieval theories of animal rationality because it shows that one can argue for the existence of an anthropological difference and at the same time leave room for continuities and similarities between different species. Thus, one could say that all later medieval authors covered in this study – or almost all, as shall be discussed in more detail below – are differentialists from a *metaphysical* point of view as all of them deny intellectual *faculties* to nonhuman animals. Still, not all of them are differentialists from an *explanatory* or *methodological* perspective because, as the following chapter will show, some chose a more assimilationist strategy for explaining certain cognitive *processes* underlying the behaviours of nonhuman animals.[28] This does not mean that their explanations are more coherent than differentialist explanations. They are simply different. But this difference has often been ignored and so it is worth bringing to the fore.

---

26 Brandom (2000), 3 [italics in the original].

27 Wild (2013), 37. Wild (2006), 9f., points out that those assimilationists who agree that there is an anthropological difference defend a *weak assimilationism*, whereas those who deny such a difference defend a *strong assimilationism*. For him, the Aristotelian model, as it was defended by Aristotle and his medieval followers, can be described as 'assimilationist account ending in a clear anthropological difference'; see ibid., 34.

28 This strategy is identical to what has been identified as the fifth and last option in Chapter 7.

CHAPTER 32

Towards a Classification: Differentialist and Assimilationist Explanations

When looking at later medieval theories of animal rationality one finds different explanatory strategies used across the board. In what follows, these different strategies shall be highlighted by taking some of the positions that have been discussed in Parts 2 to 4 concerning (I) universal cognition, (II) judging, and (III) reasoning.[1]

(I) Universal Cognition and Concept Formation

None of the authors covered in this book deny that human and nonhuman animals share a variety of cognitive capacities, most importantly, various kinds of sensory cognition. One of the main arguments for why they do so is that all animals are endowed with a sensory soul. This type of soul is what sets them apart from plants because although plants are living beings, too, they only have vegetative souls that are responsible for food intake, growth, and reproduction (see Chapter 5). In contrast to animals, plants lack perception. For instance, they neither see nor smell what they eat. Animals, by contrast, have all different kinds of perceptions. They perceive their food, other members of their species, and a great many other things. The decisive question, however, is whether there is a big difference between the way in which a human being and a dog perceive a steak, for example. Is the dog even capable of seeing a steak, that is, can it see the brownish grilled thing *as* steak? Put differently, is it capable of conceptualising its perceptual contents?

As seen in Part 2, late medieval authors discussed questions like these under the heading of what is usually called 'incidental (or accidental) perception'. They largely agreed that both humans and nonhuman animals can engage in incidental perception because it occurs whenever perceptions from different senses somehow interfere with each other. Hence, as long as an animal is endowed with more than one sensory power it is likely to experience incidental perception. It perceives something yellow as sweet or, to give another famous

1 For the sake of brevity, I will focus on the triad of intellectual operations and leave prudence out here.

 DOI 10.1163/9789004363779_040

example, it perceives something grey and furry as hostile, as the sheep does when seeing a wolf. However, different thinkers disagreed about the nature of the underlying cognitive process, more precisely, they held different views of how close nonhuman animals' incidental perceptions come to the conceptualisation of perceptual contents that we find in human beings. A good example of this disagreement is the difference between Thomas Aquinas' and Pseudo-Peter of Spain's accounts of estimative perception (see Chapters 8 and 10).

Both emphasise that the faculty of estimation plays a pivotal role in nonhuman animals' perception of the world because it is this very power that makes them perceive things *as* hostile, hurtful, friendly, useful, and so forth. Still, they give different answers to the question of whether this kind of perception is universal to some extent. For Aquinas it is clearly not. In his view, 'the estimative power does not perceive an individual insofar as it is under a common nature but only insofar as it is the starting or endpoint of some action or passion'.[2] Therefore, a sheep neither perceives a wolf as a member of the species of wolves nor does it sort wolves into the category of hostile things. It does not even possess the concept of wolf. Instead, it sees a grey, furry thing (which it might also smell or hear, for instance) and once this stimulus has passed the sheep's estimative power it immediately triggers the reaction of flight. According to Aquinas, this is exactly what the estimative power has been designed for. Its function is to guarantee the survival of an animal by triggering whatever reaction is needed in order to survive.[3] This reaction is usually hardwired and does not involve universal cognition.[4]

Now one might object that this is implausible because sheep apparently run away from *all* wolves. This seems to show that their reaction is universal and so is the cognition on which this reaction is based. However, Aquinas explains that one should not conflate the nature of the reaction with the nature of the cognitive process. Sight perceives colours universally (*universaliter*) insofar as it perceives *all* colours but it does not perceive the universal 'colour'. Similarly, the sheep by virtue of its estimative power 'hates wolves in general' (*odit lupus generaliter*). Nevertheless, its passion is not based on the cognition of

2 Thomas Aquinas, *In Aristotelis librum De anima commentarium*, lib. II, lec. 13, n. 398, ed. Pirotta (1959), 101: "Aestimativa autem non apprehendit aliquod individuum, secundum quod est sub natura communi, sed solum secundum quod est terminus aut principium alicuius actionis vel passionis; solum secundum quod est terminus aut principium alicuius actionis vel passionis."

3 Ibid.: "Naturalis enim aestimativa datur animalibus, ut per eam ordinentur in actiones proprias, vel passiones prosequendas, vel fugiendas."

4 To say that it is hardwired does not mean that some reactions of this kind are not learned, see Chapter 8.

the universal 'wolf'.[5] Its cognition does not reach beyond the level of particulars and this is where Aquinas draws the line between the perception of the world of humans and nonhuman animals. It is this difference between human and nonhuman animal cognition that he wants to illustrate by contrasting the way in which a sheep perceives a wolf and our way of perceiving things. While the sheep fails to perceive universal aspects, we are capable of cognising both particulars and universals, because in addition to various sensory faculties we have rational faculties that can engage in universal cognition. In this sense, one can say that Aquinas chooses the strategy of differentialism.[6]

A more assimilationist theory can be found in Pseudo-Peter of Spain's commentary on *De animalibus*. In his opinion, the sheep's reaction to the stimulus 'wolf' resembles a physician's reaction to a stimulus such as 'flu'. Surely, wolves that are hostile towards sheep as well as patients suffering from flu are individuals. Nonetheless, wolves are individuals of the same *kind* and so are patients with flu. In other words, all wolves belong to the same category as they share a common nature. While Aquinas holds that this common nature cannot be grasped by the sheep's estimative power, Pseudo-Peter thinks that sheep identify wolves by comparing the 'common form' (*forma communis*) of wolf with the individual they see. Although this does not mean that the sheep grasps the quiddity of wolfness its reaction of flight is based on a cognitive process that can be characterised as 'universal grasp' (*acceptio universalis*).[7] Pseudo-Peter of Spain further describes the universal nature of this cognition by introducing the concept of 'elevated intentions' (*intentiones elevatas*). Those intentions, he says, 'do not attain the same level of elevation as do those that are in the intellect'. Nonetheless, they rank above purely particular intentions. Moreover, they do not differ in kind from the universal intentions that are found in the

5 Thomas Aquinas, *Summa theologiae* I–II, q. 29, a. 6, co., ed. Leonina VI (1891), 207f.: "Potest tamen aliqua potentia sensitiva, et apprehensiva et appetitiva, ferri in aliquid universaliter. Sicut dicimus quod obiectum visus est color secundum genus, non quia visus cognoscat colorem universalem; sed quia quod color sit cognoscibilis a visu, non convenit colori inquantum est hic color, sed inquantum est color simpliciter. Sic ergo odium etiam sensitivae partis, potest respicere aliquid in universali, quia ex natura communi aliquid adversatur animali, et non solum ex eo quod est particularis, sicut lupus ovi. Unde ovis odit lupum generaliter."

6 Davids (2017), 176–178, argues that Aquinas defends a 'moderate' version of differentialism because he does not deny the many commonalities that exist between human and nonhuman animals in the realm of sensory cognition.

7 Pseudo-Peter of Spain, *Commentum super libros de animalibus* (Venetian redaction), fol. 153ra: "Alia [acceptio intentionis universalis] est prout comparat formam communem ad particulare, ut agnam ad hanc agnam; hoc non fit sine materia, et hec est anime sensitive."

intellect but they differ 'in degree' (*per gradus*).[8] Pseudo-Peter is certainly on a par with Aquinas in the sense that he denies *intellectual* universal cognition to nonhuman animals. However, he chooses the strategy of assimilationism rather than differentialism insofar as he allows for a kind of universal cognition at the level of the sensory soul, and so he assimilates human and nonhuman animal cognition in this respect.

(II) Judging

A similar difference in the choice of explanatory strategies can be found with regard to the capacity of judging. The best example is the debate between Adam Wodeham and Gregory of Rimini (see Chapter 18). In this debate, Adam Wodeham defends the view that judging is an exclusively intellectual capacity. He admits that there are cases in which nonhuman animals that lack intellectual faculties by definition seem to judge, as well. For instance, a dog standing aboard a ship mistakenly 'judges' the trees standing on the shore to be moving, and so it falls prey to an optical illusion just like we do in this situation. Yet, Adam argues that we should refrain from ascribing the capacity of judging to the dog for at least two reasons.

First, since we cannot see directly what is going on in the dog's head we should avoid anthropomorphising its cognition. From the mere fact that the dog behaves like a human being in this situation we should not conclude that its behaviour is based on the same kind of cognitive process, namely, a 'complex objective judgment' (*iudicium complexe obiective*). Instead, we should account for the dog's behaviour by referring to what Adam calls a 'simple apprehension' (*apprehensio simplex*), that is, an immediate instinctual reaction.[9] This is more concise and less anthropomorphic an explanation, hence to be preferred. The second reason why Adam opts for this explanation rather than for the ascription of judgments is that, in his view, the latter implies the ascription

8 Ibid., 152ra: "Dicendum, quod virtus sensitiva non accipit solum intentiones individuales, sed intentiones individuales elevatas. Iste tamen intentiones, quas accipit, non attingunt elevationem, que apud intellectum, sed distant per gradus."

9 Adam Wodeham, *Lectura secunda in primum librum Sententiarum*, prol., q. 4, §8, ed. Wood (1990), 99f.: "Id est non solum non libere prosequuntur ac fugiunt proficua et nociva, sed nec deliberant nec iudicant aliquid iudicio complexo obiective conveniens exsistere vel nocivum de fugiendo vel [proficuum de] prosequendo. Sed statim ad simplicem apprehensionem istius quod est nocivum fugiunt et illius quod est conveniens prosequuntur."

of 'practical reason' (*ratio practica*) to nonhuman animals.[10] This would entail a serious shift in the human/animal boundary, because the possession of reason would no longer be the mark of the human. Adam is not ready to give up this distinctive feature, therefore, he chooses the strategy of differentialism.[11] He argues that there is a crucial difference between humans and other animals to such an extent that only the former are capable of judging.

Although Adam Wodeham's opponent, Gregory of Rimini, does not deny that there are differences between human and nonhuman animals he rejects Adam's denial of the capacity of judging to nonhuman species. There are two main reasons for why he does so. The first is that he considers Adam's theory to be unsatisfactory. In his view, an explanation that refers to simple apprehensions cannot convincingly explain those cases in which animals show different reactions to like stimuli. He gives the example of an animal that sees a piece of bread. According to Adam's theory, this stimulus must always trigger the same reaction. Yet this is not what we see, Gregory objects, because sometimes we see the animal going for the bread and sometimes we see it showing no interest in the bread at all. This behaviour, Gregory thinks, cannot be explained with a simple stimulus-response model. Instead, it must be explained by referring to the capacity of judging. Every time the animal sees a piece of bread it does not immediately react to this stimulus but *judges* how it should react. The simple apprehension is thus supplemented by a judgment (*iudicium*).[12]

However, it is important to note that this kind of judgment is a sensory and not an intellectual judgment. It differs from intellectual judgments in two

10 Ibid., 99: "Sed si hoc movere deberet, esset consequenter in eis ponenda ratio practica, id est dictamen practicum de eligendis et respuendis, prosequendis et fugiendi. Sic enim agunt si dictamen haberent. Et tunc non video quare non debeant animalia rationalia appellari."

11 Cf. Perler (2006), 91f. According to Perler, Wodeham's differentialism is also motivated by the fact that he holds a theory according to which the content of judgments is a so-called '*complexe significabile*' which is "expressed by a full-fledged sentence." However, this cannot fully explain Wodeham's animosity because his opponent, Gregory of Rimini, also subscribes to this theory but still grants nonhuman animals a form of judgment that has no *complexe significabile* as its content. On the *complexe significabile* see also Nuchelmans (1973), 227–237, and (1980), 104–106.

12 Gregory of Rimini, *Lectura in primum et secundum Sententiarum*, dist. 1, q. 3, eds. Trapp & Marcolino (1981), 304: "Quod primo convinci potest, quia, sicut videmus, aliquando brutam, apprehendit aliquod sensibile, ut puta panem, et movetur ad ipsum; aliquando vero apprehendit idem, et non movetur ad ipsum. Ergo motus iste, cum sit per appetitum animalem et talis appetitus sequatur apprehensionem, praesupponit praeter simplicem apprehensionem sensibilis iudicium quo iudicatur illud utile vel necessarium aut tale vel tale."

respects. First, it does not concern universals but only singulars. The animal can judge about *this* bread but not about bread in general or the universal 'bread'. Second, it can only judge about sensible things (*sensibilia*), while humans can also judge about insensible things (*insensiblia*) such as numbers or justice.[13] It is this difference between sensory and intellectual judgments that leads to the second reason for Gregory's rejection of Adam's view. For Adam the ascription of judgments implies the ascription of practical reason and a shift in the animal/human boundary. For Gregory this is not the case. He thinks that one can *assimilate* human and nonhuman animals without giving up the denial of intellectual faculties to the latter. Humans judge about all different kinds of things: particulars, universals, sensible, and insensible things, and they do this because they are endowed with both sensory and intellectual powers. Nonhuman animals also judge but they lack intellect and reason and so their judgments are restricted to the realm of particular, sensible objects. Still, they engage in some form of judging and despite all differences this establishes an important cognitive *similarity* between humans and other animal species. In contrast to Adam Wodeham, Gregory of Rimini thus chooses assimilationism instead of differentialism.

(III) Reasoning

The third element of the triad of intellectual operations is reasoning. Just like the first and second element, universal cognition and judging, this element gave rise to the question of whether or not it belongs to human animals alone. There are many situations in which nonhuman animals are confronted with a certain problem and, from the outside, seem solve this problem in the same way that we do. One example that was debated by late medieval authors in this context is this: a dog has the choice between a longer and a shorter path in order to reach its destination. As long as nothing blocks the shorter path it will choose this option simply because it is the best option available. However, it will choose the longer path when the shorter path is blocked and so it takes a detour for a very good reason. The question now is whether or not its

---

13 Ibid., 305: "Nec valet ratio cuiusdam doctoris, qui dicit quod secundum hoc in talibus animalibus esset intellectus practicus et deberent talia animalia appellari rationabilia. Non enim haec sequuntur loquendo de huiusmodi ad sensum, in quo dicimus hominem esse rationale et habere practicum intellectum, quia homo non tantum singularia iudicia sed et universalia, nec tantum circa sensibilia sed et circa insensibilia habet iudicia, et ex talibus ratiocinatur et proprie rationalis appellatur."

behaviour is actually based on a syllogistic process. Does it employ a syllogism of the form 'The shorter path is to be favoured over the longer path because it is faster but now that it is blocked I must take the longer path even though this will take more time'?

A negative answer to this question was given by John Duns Scotus (Chapter 19). He argues that from similarities in behaviour we should not infer to 'similar cognition' (*cognitio similis*).[14] Instead, we should examine whether there could also be different cognitive processes leading to the same behavioural results. In the case of the dog that makes a choice between two paths the behaviour derives 'from sensory appetite alone' (*ex solo appetitu sensitivo*), Scotus argues. By contrast, the behaviour of a human being who is confronted with the same problem originates from 'rational discourse' (*discursus rationis*) and 'deliberation' (*deliberatio*). Unsurprisingly, the main reason for this difference is the metaphysical difference between dogs' sensory and humans' rational souls. So even though dogs and human beings show the very same behaviour in this situation, the cognitive processes underlying this behaviour are essentially different. Those processes do of course lead to very similar behavioural outcomes which is why they can be characterised as *functionally equivalent*. This means that the function they have is basically the same. Still, the way in which this function is realised, namely, by virtue of a particular process is fundamentally different.[15]

Obviously, Scotus *differentiates* between human and nonhuman animals in this respect as he argues that only humans can engage in rational thinking or reasoning. For him a dog is not very much different from a cleverly constructed machine such as a calculator, one could say. The calculator will easily solve numerous mathematical problems, and it often does this many times quicker than we do. Nevertheless, we are relatively reluctant to say that the calculator is more intelligent than we are or that it thinks and reasons. Instead, we say that it is a machine that has been designed by intelligent, thinking beings for the purpose of automatically solving certain tasks quite quickly, much like the chess-playing computer that is famously mentioned by Daniel Dennett or the

14 See John Duns Scotus, *Quaestiones super libros Metaphysicorum Aristotelis*, lib. 1, q. 3, §§10f., eds. Andrews et al. (1997), 89: "Bruta multa faciunt ex cognitione eodem modo quo fierent ab homine cognoscente per discursum rationis [...]. Ad hoc dicitur quod licet similiter agant quaedam, sicut homo ex deliberatione ageret, non tamen oportet quod similem cognitionem habeant. Nam illud, quod eligitur ex deliberatione, posset etiam idem non eligi ex deliberatione, sed ex solo appetitu sensitivo."

15 The concept of functional equivalence now plays an important role in the debate over machine intelligence and morality; see Wallach & Allen (2009), 68.

heat-seeking missile that features so prominently in Donald Davidson' paper on rational animals and Fred Dretske's book on behaviour.[16] The point of these examples is that if we claim that the dog is reasoning in this situation we fall into the trap of anthropomorphism, making the dog's cognition much more humanlike than it actually is.[17] Instead, we should only call it 'quasi-rational', for example.[18] This warning is typical for differentialists and Scotus but also Thomas Aquinas can be read in this way.

Some authors, however, preferred a different explanation. A prominent example is John Buridan whose explanation of the dog's cognition is the exact opposite of Scotus' theory (see Chapter 22). For Buridan there is no reason to assume that the dog is incapable of reasoning. Rather, there are very good reasons to hold the contrary, because the souls of humans might simply be as material as the souls of dogs, donkeys, and other nonhuman animals. To those who reject this idea 'because of the great subtlety of humans or their reasoning' (*propter magnam subtilitatem hominis vel eius ratiocinationem*) Buridan replies that reasoning might simply be a matter of material complexity and complexion. Therefore, a dog that has to make a choice between two paths, one of which is short but blocked, will take the longer path on the basis of reasoning.[19] Buridan does of course pay a metaphysical price for this position: he gives up the distinction between the immaterial rational and the material sensory soul.[20] In this sense, his assimilationism is not only *methodological* but also *metaphysical*. However, one should not ignore that Buridan still stands in the tradition of the Aristotelian denial of reason because, like other late medieval thinkers, he does not explicitly ascribe the faculties of intellect and reason

16 See Dennett (1971); Davidson (1982), 323f.; Dretske (1988), 126. On thinking machines see also Searle (2002).

17 On the problem of anthropomorphism see Wild (2013), 69–73; De Waal (1999); Daston & Mitman (2005); Wynne (2007).

18 While Scotus does not explicitly employ this terminology, Thomas Aquinas, for instance, says that the dog behaves '*as if* using a disjunctive syllogism' (*quasi utens syllogismo divisivo*); see p. 135 n5.

19 John Buridan, *Quaestiones de anima*, Paris, Bibl. Nat., Cod. lat. 15888, f. 70ra: "[...] canes et alia animalia ratiocinantur et syllogizant, quamvis non ita subtiliter ac complete sicut homo vel simia. Quod apparet, quia, si canis videt dominum suum et vult ire ad ipsum et in directa linea inveniat magnam foveam, non intrabit in illam, sed quaerit aliam viam, licet longiorem, quod non faceret, nisi ratiocinaretur et syllogizaret, quod non est bonum cadere in foveam et cetera."

20 A Renaissance thinker who was also willing to pay this price is Pietro Pomponazzi; see Rubini (2015).

to nonhuman animals. Rather, he questions, if not rejects, the traditional view according to which certain cognitive processes, such as reasoning, cannot be had without such faculties.

Now one might think that this view always comes at the price of giving up the metaphysical distinction between different types of souls. Yet as the example of Albertus Magnus shows this is not necessarily the case. Like Buridan, Albert claims that nonhuman animals are not per se incapable of reasoning (Chapter 21). There are at least some highly-developed species, such as apes (to which belongs the subspecies of pygmies), that can engage in what Albert calls 'imperfect practical syllogisms' (*imperfecti sillogismi operum*).[21] The pygmy in particular also 'seems to somehow participate in making inductive inferences' (*inductionis autem aliquid videtur participare*), Albert states, but he also notes that it does not do this perfectly because it lacks cognitive access to universals (see Chapter 9). Consequently, its inferences remain imperfect in comparison to ours. One could perhaps compare the reasoning of pygmies to the reasoning of human children. They make all sorts of inferences but their reasoning does not operate as smoothly as the reasoning of an adult human being and, above all, is less complex.

Be that as it may, the crucial point here is that reasoning is no longer the prerogative of the rational soul. There are certain basic types of reasoning in which nonhuman animals can engage by virtue of their sensory powers. One particular power that Albert highlights in this respect is the faculty of estimation. The estimative power of apes, he explains, is 'more lucid' (*lucidior*) than the estimation of other nonhuman animals and even amounts to something that is 'similar to reason' (*rationi simile*).[22] So, unlike Buridan, he maintains the metaphysical distinction between immaterial rational souls and material sensory souls but at the same time 'upgrades' the latter, especially the estimative power. This means that he ascribes to those sensory faculties certain capacities, such as reasoning, that were usually considered to require an immaterial faculty. Now one might say that this claim is rather unconvincing because it ignores the link between the *immateriality of a faculty* and the *rationality of its operations*.

21 Albertus Magnus, *De animalibus*, lib. XXI, tr. 1, c. 3, ed. Stadler (1920), 1331: "[…] ita in istis animalibus sunt imperfecti sillogismi operum qui non habent nisi fantasticam aestimationem de operabili vel appetibili et appetitum facientem impetum ad opus."

22 Ibid., 1332: "[…] et ideo pro certo aestimativam haec animalia habent lucidiorem omnibus aliis animalibus: propter quod melius aestimant de fantasmatibus quam alia animalia et melius eliciunt intentiones acceptas cum fantasmatibus quam aliqua aliorum animalium, ita quod videntur aliquid habere rationi simile."

Yet what the assimilationist approaches show is exactly this: the link between the nature of a faculty and the nature of its operations can be loosened at certain points, and so can the link between faculty rationality and process rationality. In this sense, they create room for rationality at the level of the sensory soul and develop the idea of what might be called 'rationality without reason'.

CHAPTER 33

Room for Rationality or Rationality without Reason

According to the standard account of medieval theories of animal cognition, human and nonhuman animals differ in at least one crucial regard: rationality. While humans have it, nonhuman animals lack it. There are pretty strong reasons to believe that this account is correct. Historically, there is the success of Aristotelian faculty psychology. It was quite successful in Antiquity and, with the transmission of Aristotle's writings, encountered success in the later Middle Ages, too. Briefly speaking, faculty psychology is the idea that cognitive (and also certain non-cognitive) *processes* depend on *faculties*. What this means for rationality is this: rational processes such as concept formation, judging, or reasoning rely on rational faculties. To put it in modern terms, *process rationality* cannot be had without *faculty rationality* because the lack of rational faculties causes the lack of rational processes.

By and large, there is not much to say against this account because it simply summarises the Aristotelian background of later medieval theories. The Aristotelian model of the soul is not only the starting point of those theories but also the framework within which they are supposed to work. Yet a finding of this study is that this account leaves out some important pieces of the picture. One of those pieces is that the Aristotelian denial of reason or the metaphysical idea of faculty rationality did not lead into an explanatory one-way road. When it came to explaining the cognitive processes that underlie the behaviours of different animals authors employed different explanatory strategies which can be classified as *differentialist* and *assimilationist*. This means that some of them were keen to stress the differences that exist between human and nonhuman cognition, while others put more weight on the similarities. Most, if not all, of them did, of course, defend some version of *metaphysical differentialism* insofar as they stuck to the Aristotelian denial of reason in terms of *faculty rationality*. Nonetheless, some chose what could be called an *explanatory assimilationism*. So, while denying rational *faculties* to nonhuman animals, on the one hand, they ascribed them certain rational *processes*, on the other hand.

This kind of explanatory assimilationism leads to another interesting piece of information that is usually missing from the standard account, which can be called *rationality without reason*. What it means is basically this: in some cases nonhuman animals can engage in rational processes although they lack the

 | DOI 10.1163/9789004363779_041

correspondent faculties. We have come across some examples of rationality without reason in the previous chapter and among the more obvious cases surely is the ascription of reasoning to nonhuman animals as found in Albertus Magnus and John Buridan. One could also add Nicole Oresme to this list, as well as some of those views concerning other parts of the triad of intellectual operations and prudence, such as Pseudo-Peter of Spain's account of universal cognition and Gregory of Rimini's account of judging. In all of these cases, the rationality ascribed to nonhuman animals is not just any kind of rationality. It differs from the full-fledged rationality that is usually ascribed to humans in at least three regards: (I) it is *rational*, not intellectual, (II) *practical*, not theoretical, and (III) *particular*, not universal.

(I) Rational, Not Intellectual

Throughout this book the terms 'intellect' and 'reason' have been used quite interchangeably. This is very much in line with the terminology of the thinkers covered in this study, because in denying rational faculties to nonhuman animals they did not differentiate between intellect and reason. This does not mean, however, that they ignored the differences that exist (or might exist) between these powers. As mentioned in Chapter 6, Thomas Aquinas, for instance, discusses in quite some detail whether or not they are separate powers.[1] His ultimate answer is no, but he points out that there is a difference between the corresponding operations, between 'reasoning' (*ratiocinari*) and 'understanding' (*intellegere*).[2] While '*intellegere*' means to understand something directly (*simpliciter*), reasoning is a more indirect way that involves various steps until the point of understanding is reached. This more or less subtle difference has so far not mattered much. In the present context, however, it becomes relevant, because if late medieval authors attributed rationality to nonhuman animals at all it was in terms of '*ratiocinari*' rather than of '*intellegere*'. John Buridan, for instance, claims that dogs and apes are capable of 'reasoning' (*ratiocinari*) and 'syllogising' (*syllogizari*) but he does not employ the term '*intellegere*' in this context.[3] Similarly, Nicole Oresme argues that they can mentally 'run

1 See Thomas Aquinas, *Summa theologiae* I, q. 79, a. 8, ed. Leonina V (1889), 274f.

2 '*Intellegere*' is commonly translated with 'to understand sth'. but one could also translate it with 'to grasp sth. intellectually'.

3 See p. 224 n19.

through' something (*discurrere*) but this, too, is a rational, not an intellectual process, insofar as it means to understand something step-by-step rather than all at once.[4] One could also mention Roger Bacon in this context because even though he argues that it is rather misleading to ascribe 'cognition by syllogism' (*cognitio per sillogismum*) to other animals, he still thinks that they can engage in some sort of mental 'discourse' (*decursus*; *discursus*).[5] He therefore ascribes them the faculty of cogitation which was usually taken to belong to the rational part of the soul. In sum, what this shows is that if there is rationality without reason at all, it is – somehow tautologically – rationality in a rational, not in an intellectual sense of the term.

(II) Practical, Not Theoretical

A second important feature of rationality without reason is its practical nature. This means that such animals can reason about what is practically relevant to them, but not about more theoretical issues. For instance, a monkey might reason that it should eat this yellow curved thing because another yellow thing that looked very similar and that it ate some time ago was quite delicious. In this case, the monkey's reasoning concerns a practical issue: shall I eat or shall I not eat this? Its reasoning does not concern theoretical questions such as 'Why do some philosophers think that the specific taste I experience when eating a yellow curved thing is a so-called quale?' The monkey simply does not care about qualia (even though it might have them) because they are practically irrelevant for its decision about eating or not eating. What matters are those questions that are immediately relevant for its survival, that is, what some people like to call 'the four Fs' of evolutionary biology: fighting, fleeing, feeding, and mating. The practical nature of this kind of reasoning is made explicit by Albertus Magnus who claims that apes are capable of employing 'imperfect practical syllogisms' (*imperfecti sillogismi operum*).[6] Similarly, Gregory of Rimini emphasises that nonhuman animals do not judge about 'insensible things' (*insensibilia*). To be clear, this does not mean that humans' rationality, in contrast, is entirely theoretical. Rather, it is *also* theoretical, because we can reason and judge about both practical and theoretical matters.

4 Nicole Oresme, *Expositio et quaestiones in Aristotelis De anima*, lib. III, q. 4, ed. Patar (1995), 335.

5 Roger Bacon, *Perspectiva*, pars II, dist. 3, c. 9, ed. Lindberg (1996), 250.

6 See p. 225 n21.

(III) Particular, Not Universal

The third and final feature is that the rational processes nonhuman animals can engage in do not involve universals. To take up the example from before, the monkey does not possess the concept of banana. It reasons about what it perceives at the moment, such as 'this yellow curved thing', but not about bananas or yellow curved things in general. Surely, someone like Pseudo-Peter of Spain might intervene and say that many nonhuman animals can of course grasp certain universal aspects of things. Still, even he admits that this kind of universal grasp does not attain the level of intellectual concept formation and universal cognition one finds in human beings. Albertus Magnus therefore says that if nonhuman animals reason at all, they reason about what is 'here and now' (*hic et nunc*) and what is here and now is something singular, not universal.[7] Their rationality is thus also what is called 'context-dependent' in the modern debate.

Now if one looks at these qualifications, it becomes clear that there are important differences between nonhuman animals' rationality without reason and human rationality with reason. Human rationality is not only rational but also intellectual, it is not only practical but also theoretical, and it is particular as well as universal. Nevertheless, there is some common ground, something both human and nonhuman animals share, and this in some sense runs counter to the traditional rationality/non-rationality divide. Rather than just one type of rationality that is largely based on faculty rationality there are different types or levels of rationality and this is exactly what is also stressed by contemporary theories of animal rationality. They do not aim to put nonhuman animals on the same level as humans. Rather, they try to discover commonalities *and* differences and what is interesting to see is that there is a certain continuity between later medieval and contemporary theories in this respect. Most contemporary scholars also emphasise that the process rationality we find in nonhuman animals is practical, not theoretical, particular, not universal (or *domain-specific*, not *domain-general*, as they put it).[8] So even though they have left behind the framework of Aristotelian faculty psychology they keep making very similar qualifications when it comes to the question of rationality in nonhuman animals.

7 Albertus Magnus, *De animalibus*, lib. XXI, tr. 1, c. 3, ed. Stadler (1920), 1331.

8 See Hurley & Nudds (2006b), 13–19. See also Bermúdez (2003), 38–46, who presents what he calls "minimalist approaches to non-linguistic thought." These accounts are characterised by four features: (i) context-bound, (ii) pragmatic, (iii) sensory, (iv) unstructured; see ibid., 45.

However, one could object to this analysis that the idea of rationality without reason or process rationality without faculty rationality is simply incompatible with Aristotelian psychology. One might say that no serious Aristotelian would claim that there can be whatever process *without* the corresponding faculty. This is absolutely correct, but misses the actual point. The point is not that the idea of rationality without reason makes Aristotelian faculty psychology collapse. None of the authors covered in this book would deny that every process relies on a faculty. What some of them seem to suggest is simply that not all of those processes that were usually taken to depend on rational faculties, actually require those faculties. Again, this does not mean that they do not require any faculties at all. Rather, they do not require the faculties of intellect and reason but can take place in the sensory soul (or some part of the sensory soul). Therefore, one could say that the sensory soul is 'upgraded' in the sense that the scope of its cognitive operations is widened.

It is important to note that this widening does not make the rational faculties superfluous. As we have seen, the reasoning, for example, in which the sensory faculties such as the pygmies' estimative faculty might engage is only a very basic form of reasoning. Similarly, sensory judgments are very simple judgments that lack the predicative structure and complexity of intellectual judgments. Moreover, it should be noted that quite often philosophers do not consistently argue for such an expansion of the sensory soul's capacities. For instance, Buridan can be interpreted as a differentialist on concept formation but as an assimilationist on reasoning. Gregory of Rimini presents a differentialist explanation concerning reasoning, but gives assimilationist arguments with regard to judging. And Albertus Magnus seems to choose the strategy of differentialism as far as universal cognition is concerned, but assimilationism with regard to reasoning and prudence. This inconsistency arises partly from the peculiar conditions under which medieval theories of animal rationality were developed (see Chapters 4 and 30). As has been shown, they were in many cases nothing but a by-product of a debate that was not primarily concerned with animals and their cognition.

Despite this historical peculiarity, the assimilationist explanations that one finds in the later Middle Ages establish an interesting parallel to both ancient and early modern theories. Already in Antiquity, "the denial of reason and belief to animals necessitated an expansion of their perceptual content," as Richard Sorabji remarks.[9] This expansion became even more common with the growing success of the theory of inner senses in the late ancient and early medieval period because the inner senses, most importantly imagination and

9 Sorabji (1993b), 20.

estimation, have the potential to account for various complex behaviours in all kinds of animals. A very similar strategy can be found in the early modern period. An illuminating example is Marin Cureau de la Chambre who in his *Traité de la connoissance des animaux*, published in 1648, argues that nonhuman animals can engage in 'judging' (*jugement*) and 'practical reasoning' (*raissonement*) although they lack the faculties of intellect and reason.[10] The explanation Cureau gives is that the imagination and other sensory powers enable them to do this. So he upgrades those powers by ascribing them certain capacities that were traditionally taken to belong to the higher faculties, namely, intellect and reason.[11]

Still, like the later medieval Aristotelians we have examined, Cureau emphasises that the reasoning of nonhuman animals is inferior to the inferential capacities of human beings, first and foremost, because it does not involve universals.[12] He thus defends what is called 'abstractionism'. This is the thesis that nonhuman and human animals share a wide range of cognitive capacities except for the ability to abstract from the level of particulars to the extent that one can form universal concepts. In the abstractionist theory, reason is replaced by concept formation as anthropological difference and so concept formation and universal cognition establish a cognitive boundary between human and nonhuman animals. In addition, there also remains a metaphysical boundary. According to Cureau, only human souls are characterised by 'spirituality and immortality' (*spiritualité et immortalité*).[13] So, on the one hand, he assimilates human and nonhuman animals by ascribing to them a wide range of shared capacities such as judging and reasoning. At the heart of this lies the idea that sensory powers are capable of much more than usually thought. On the other hand, they cannot attain the level of intellectual powers in each and every respect. Therefore, one must not ignore the differences that exist.[14]

Cureau's position very much resembles several later medieval theories of animal rationality. The strength of the explanation of animal rationality that they give is that it allows us to maintain the framework of Aristotelian psychology.

10 Marin Cureau de la Chambre, *Traité de la connoissance des animaux* (Paris 1648), 5.

11 On his position see Wild (2006), 19–21.

12 Cureau de la Chambre, *Traité de la connoissance des animaux*, 5f. and 9. See also Wild (2006), 20f.

13 Cureau de la Chambre, *Traité de la connoissance des animaux*, 9.

14 Therefore, Wild (2006), 10, classifies Cureau's assimilationism as '*weak* assimilationism'. Other early modern defenders of such a weak assimilationism are Locke and Leibniz; see ibid., 290. Hume ultimately defends a strong assimilationism by claiming that there is no anthropological difference at all. In Hume's view, reason is not a separate faculty but simply a function of the power of imagination, see ibid., 264.

The basic idea that processes rely on faculties remains intact. What is slightly modified is the correspondence between certain rational processes and rational faculties. Although this shift is often ignored one should not overestimate its role. It is certainly exaggerated to say that the "principle of a qualitative difference between humans and animals" that had been established by Aristotle and the early church fathers began to break down by the later Middle Ages, as Joyce E. Salisbury, for instance, claims.[15] Surely, the rediscovery of Aristotle's writings on psychology, zoology, natural philosophy, and other subjects gave a major boost to discussions of this principle. Nevertheless, it did not break down in the sense that authors began to ascribe intellect and reason to nonhuman animals and the holding on to this particular definition of the anthropological difference truly is an important difference to today's discussion.

One might ask why this happened. One of the most obvious explanations is that, from the viewpoint of medieval Christian theology, "the difference between human beings and brute animals is of course the difference that matters, since it is we who are made in God's image and to whom eternal life has been promised," as Jack Zupko rightly notes.[16] So there was a *theological* interest in maintaining man's special status, his status as 'most excellent animal' (*animal nobilissimum*).[17] Still, the later medieval period saw a lively *philosophical* discussion on the question of what this special status amounts to. As shown, there was a whole spectrum of answers available and within this spectrum one can also find the view that the denial of reason does not necessarily imply the denial of rationality.

15 Salisbury (2011), 1–3.
16 Zupko (2008), 184.
17 See Köhler (2008) and (2014). For a critical evaluation of this concept see Anzulewicz (2013a) and (2016).

Conclusion

This book began with the story about a crisis – the crisis that, according to Richard Sorabji, was produced "in psychology, in ethics, and in religion" by Aristotle's denial of reason to nonhuman animals.[1] There is no doubt that Aristotle did indeed grant the faculties of intellect and reason to human beings alone, as did his late medieval followers. The crucial question, however, is whether the consequences of this denial were as critical as Sorabji has famously suggested. Did it really lead to an excessive downgrading of nonhuman animals and a valorisation of humans, especially in the medieval period which was and still is described as a 'dark age' in this regard?

As has been shown, one cannot give a straightforward answer to this question. But as far as the field of psychology is concerned, it has become clear that within the Aristotelian theory of the soul there are certain *grey areas*. There is, on the one hand, a clear distinction between humans and other animals: while the former possess rational or intellectual souls, the latter have sensory souls. Consequently, the behaviours of nonhuman animals are explained by the cognitive processes that take place in the external and internal senses, most of which they share with humans. Human behaviours, in contrast, are accounted for by the operations of the senses but also, and perhaps even mostly, by those operations that originate from the faculties of intellect and reason, most importantly the cognition of universals and the formation of concepts, judging, and reasoning. It is these cognitive operations or capacities through which humans excel other animals, according to this theory.

On the other hand, this clear-cut distinction gave rise to a variety of questions. One of these questions was whether one can really explain all kinds of nonhuman animal behaviours in a satisfying fashion without referring to rational operations. Various species show behaviours that strikingly resemble our own, and therefore it seems as if they possess the same cognitive powers we have, since similar effects seem to have similar causes. The answers that were given to this and other questions by thirteen thinkers between roughly 1250 and 1350 were the subject of this book, and they have been called *theories of animal rationality*. If one takes this to refer to both human and nonhuman animals it might seem like a clear contradiction in terms because none of the authors covered here actually ascribed rational *faculties* to nonhuman animals. Nonetheless, all of them seriously discussed the strengths and weaknesses of this denial and some even left some room for what could be called 'rationality

1 Sorabji (1993a), 1.

 | DOI 10.1163/9789004363779_042

without reason'. This means that some of them, for instance, concluded that certain highly developed species of nonhuman animals can engage in rational *processes*, such as basic forms of reasoning although they lack intellect and reason. To be clear, this was not the majority view. By and large, late medieval authors stuck to the line which had been drawn by Aristotle. But they admitted that this line is blurred in some places, and they realised that it is not always easy to define or redefine its course: nonhuman animals, for example, sometimes seem to have general mental representations, and they also seem to judge about what they perceive. Surely, this does not mean that they form concepts and propositions but their cognitive life is still much richer and much more complex than the denial of reason suggests. This complexity did not escape the attention of medieval scholars, and they chose different explanatory strategies for dealing with it. It is this dimension or this spectrum of explanations that the story about the crisis produced by Aristotle risks obscuring.

Still, one should not discard claims about this crisis too quickly. Rather, one should further examine the denial of reason and its implications. In this respect, one of the most interesting findings of this book is that, despite much sensitivity for the complex cognitive life of nonhuman animals in the Middle Ages, the anthropological difference remained untouched. Although various authors questioned the denial of rational faculties to nonhuman animals, none of them ultimately rejected it. Instead, they tried to formulate theories that coherently account for nonhuman animal behaviour without such faculties. Many of their theories significantly differed from each other, as has been shown. While some held that natural instinct can explain most of what we see in other animals, others claimed that certain high-level cognitive operations can be performed without higher cognitive faculties. But even in this latter case, the denial of reason remains. Hence, it would be worth asking why they did not go beyond this point? Did they fear that the ascription of reason and intellect to other animals would seriously endanger man's exceptional position? And if so, was this fear motivated by the Christian doctrine of the immortality of the soul and other doctrines? How did these affect or interfere with the philosophical discourse in the Middle Ages, especially in the case of animal rationality? This is a promising field for further enquiry.

Other fields of potentially illuminating enquiry include the emotional and volitional life of nonhuman animals. The present book has focused mainly on the aspect of cognition and even more narrowly on higher-level cognitive capacities. But there is more to say about animal minds because both human and nonhuman animals seem to *feel* and to *want* things. The question, however, is how the lack or the possession of rational faculties affects emotions and volitions. One could, for instance, ask whether those animals that lack intellect and reason can control their feelings and desires. Do they have different emotions than we do? And are they actually capable of *wanting* something at all since the will was usually taken to be a power of the rational soul? Some of these questions have already been addressed in current scholarship, but there still remains much to investigate in order to render a more complete picture of the medieval *philosophy of animal minds*.[2] Part of this picture would also include other aspects, such as the questions of consciousness, self-awareness, and self-knowledge.[3] On the one hand, these aspects belong to the field of cognition but, on the other hand, there might also be an awareness of feelings and desires in nonhuman animals, and so these questions actually concern the mental life of other animals in general.

The field of animal minds is, furthermore, linked to another field, namely, the field of animal ethics.[4] This connection is not an originally modern one, as several studies on animal ethics in ancient and early-modern times have shown.[5] The moral status of animals often, if not always, depends on their metaphysical status and this, in turn, is highly dependent on their mental status and on their cognitive capacities, in particular because the cognitive capacities often establish a criterion to decide whether or not a being should be treated as a member of the moral community. This was also the case in the Middle Ages. However, a comprehensive study on animal ethics in the medieval period is still missing.[6] An interesting question for such a study would be whether the *metaphysical differentialism* of medieval thinkers also produced a *moral differentialism*. Moral differentialism is the view that there is a fundamental difference between us and other animals – an *anthropological difference* – and this difference has consequences for our treatment of nonhuman animals.[7] As the denial of reason establishes such a difference, the question would thus be how it affects the definition of moral differences. Did medieval philosophers completely deny any kind of moral rights to nonhuman animals or did they

---

2 On animal emotions in medieval philosophy see Loughlin (2001); Perler (2012a) and (2012b); Köhler (2014), 659–687. Some aspects are also covered by Brungs (2002); Miner (2009); Perler (2011); Davids (2017), esp. 202–212; Schäfer & Thurner (2013). On animal volitions see Frampton (2008); Toivanen (2011) and Köhler (2014), esp. 561–612.

3 On these aspects in Islamic philosophy see López-Farjeat (2012), (2013) and (2016). On medieval Latin views see Toivanen (2013a) and (2013b); Perler & Schierbaum (2014).

4 On this connection in current philosophy see the contributions to Petrus & Wild (2013), esp. Glock (2013). For a concise introduction to animal ethics see Grimm & Wild (2016).

5 See, for instance, Sorabji (1993b); Steiner (2005) and (2008); Gilhus (2006); Newmeyr (2011); Harden (2013); Muratori & Dohm (2013).

6 An overview of animal ethics in the Middle Ages is provided by Oelze (forthcoming a).

7 On moral differentialism see Grimm & Wild (2016), 32–40.

acknowledge that other animals are not only moral patients but moral agents, as well? In this context, one could also ask how the phenomenon of animal trials is to be understood because in this case nonhuman animals seem to be treated as morally responsible subjects. In short, medieval animal ethics is another promising field for further investigation.

The present study thus shed light on only one aspect among many others. But this aspect is crucial because it concerns the very definition of rationality, which provides the foundation for defining the animal/human boundary. Surely, this definition has changed over the centuries. And so later medieval theories of animal rationality might look strange from our perspective, not only because they were developed in a very different historical context but also because they started from a very strong anthropological premise which is no longer so widely shared. However, they also look very familiar in at least one crucial regard. If one asks a contemporary philosopher what she takes to be her job in the current debate over animal minds, she will likely answer that her task consists “not in collecting new empirical data about animal behavior, its neurological causes or its evolutionary origins, but in clarifying what it is to possess various mental properties, and hence in clarifying under what conditions such properties can be ascribed to organisms,” as Hans-Johann Glock put it.[8] Medieval philosophers might just have given the same answer. And so even if one does not share their views, one can hardly deny that they did fulfil their philosophical responsibility.

---

8 Glock (1999), 174.

Bibliography

Primary Sources

Adam Wodeham, *Lectura secunda in librum primum Sententiarum*, ed. Rega Wood (St. Bonaventure, NY: Franciscan Institute, 1990).

Adelard of Bath, *Conversations with his Nephew. On the Same and the Different, Questions on Natural Science and On Birds*, ed. Charles Burnett (Cambridge: Cambridge University Press, 1998).

Albertus Magnus, *De animalibus libri XXVI*, 2 vols., ed. Hermann Stadler (Münster: Aschendorff, 1916–20).

Albertus Magnus, *Liber de natura et origine animae* (Opera omnia 12, Ed. Colon.), ed. Bernhard Geyer (Münster: Aschendorff, 1955).

Albertus Magnus, *Quaestiones super De animalibus* (Opera omnia 12, Ed. Colon.), ed. Ephrem Filthaut (Münster: Aschendorff, 1955).

Albertus Magnus, *Metaphysica. Libri quinque priores* (Opera omnia 16/1, Ed. Colon.), ed. Bernhard Geyer (Münster: Aschendorff, 1960).

Albertus Magnus, *De anima* (Opera omnia 7/1, Ed. Colon.), ed. Clemens Stroick (Münster: Aschendorff, 1968).

Albertus Magnus, *Super Ethica commentum et quaestiones* (Opera omnia 14/2, Ed. Colon.), ed. Wilhelm Kübel (Münster: Aschendorff, 1987).

Albertus Magnus, *De homine* (Opera omnia 27/2, Ed. Colon.), eds. Henryk Anzulewicz and Joachim R. Söder (Münster: Aschendorff, 2008).

Albertus Magnus, *De memoria et reminiscentia* (Opera omnia 7/2A, Ed. Colon.), ed. Silvia Donati (Münster: Aschendorff, 2017).

Alhacen, *Alhacen's Theory of Visual Perception. A Critical Edition, with English Translation and Commentary, of the First Three Books of Alhacen's* De aspectibus, *the Medieval Latin Version of Ibn al-Haytham's Kitāb al-Manāẓir*, 2 vols., ed. A. Mark Smith (Philadelphia: American Philosophical Society, 2001).

Ambrose of Milan, *Hexaemeron libri sex* (PL 14, 133–288), ed. Jacques-Paul Migne (Paris 1882).

Anonymous, *De anima et potenciis eius*, ed. René A. Gauthier, *Revue des Sciences philosophiques et théologiques* 66 (1982), 3–55.

Aristoteles Latinus, *Metaphysica, Lib. I-XIV, Recensio et translatio Guillelmi de Moerbeka* (AL XXV 3,2), ed. Gudrun Vuillemin-Diem (Leiden/New York/Köln: Brill, 1995).

Aristoteles Latinus, *De progressu animalium. De motu animalium. Translatio Guillelmi de Morbeka* (AL XVII 2,2/3), ed. Pieter De Leemans (Brepols: Turnhout, 2011).

Aristotle, *Physica*, ed. William D. Ross (Oxford: Clarendon Press, 1950).

Aristotle, *Metaphysica*, ed. Werner Jaeger (Oxford: Clarendon Press, 1957).

Aristotle, *De anima*, ed. William D. Ross (Oxford: Clarendon Press, 1961).

Aristotle, *Parts of Animals*, ed. and tr. Arthur L. Peck (Cambridge, Mass./London: Harvard University Press, 1968).

Aristotle, *Politica*, ed. William D. Ross (Oxford: Clarendon Press, 1997).

Aristotle, *Movement of Animals*, ed. and tr. Edward S. Forster (Cambridge, Mass./London: Harvard University Press, 2006).

Les Auctoritates Aristotelis. Un florilège médiéval. Étude historique et édition critique, ed. Jacqueline Hamesse (Leuven: Publications Universitaires, 1974).

Augustine, *De Genesi ad litteram libri XII* (PL 34, 245–486), ed. Jacques-Paul Migne (Paris 1887).

Augustine, *De libero arbitrio. Libri tres* (Corpus Christianorum Series Latina 29), ed. William M. Green (Turnhout: Brepols, 1970).

Avicenna Latinus, *Liber de anima seu Sextus de naturalibus*, 2 vols., ed. Simone van Riet (Louvain: Peeters/Leiden: Brill, 1968–72).

Basil of Caesarea, *Homiliae IX in Hexaemeron* (PG 29, 1–208), ed. Jacques-Paul Migne (Paris 1886).

Bonaventure, *Commentarius in secundum librum Sententiarum* (Opera omnia II), ed. Collegium S. Bonaventurae (Quaracchi: Collegium S. Bonaventurae, 1885).

Giles of Rome, *Questiones metaphisicales* (Venice 1501).

Gregory of Rimini, *Lectura super primum et secundum Sententiarum*, eds. A. Damasus Trapp and Venicio Marcolino (Berlin/New York: Walter de Gruyter, 1981).

Henry of Ghent, *Quodlibet I*, ed. Raymond Macken (Leuven: University Press / Leiden: Brill, 1979).

Isidor of Seville, *Étymologies XII. Des animaux*, ed. Jacques André (Paris: Belles Lettres, 1986).

John Blund, *Treatise on the Soul*, eds. Daniel A. Callus and Richard W. Hunt, tr. Michael W. Dunne (Oxford: Oxford University Press, 2013).

John Buridan, *In Metaphysicen Aristotelis Quaestiones* (Paris 1518; repr. Frankfurt am Main: Minerva, 1964).

John Buridan, "Jean Buridan's *De motibus animalium*," eds. Frederick Scott and Herman Shapiro, *Isis* 58/4 (1967), 533–552.

John Buridan, *Summulae de demonstrationibus*, ed. Lambert M. de Rijk (Groningen: Ingenium, 2001).

John Buridan, *Quaestiones super octo libros physicorum Aristotelis (secundum ultimam lecturam), Libri I-II*, eds. Michiel Streijger and Paul J.J.M. Bakker (Leiden/Boston: Brill, 2015).

John Buridan, *Quaestiones in Analytica Priora*, ed. Hubert Hubien [unpublished typescript; undated].

John Duns Scotus, *Quaestiones super libros Metaphysicorum Aristotelis I–V* (Opera philosophica 3), eds. Robert Andrews et al. (St. Bonaventure, NY: The Franciscan Institute, 1997).

John Duns Scotus, *Quaestiones super secundum et tertium De anima* (Opera philosophica 5), eds. Carlos Bazán et al. (St. Bonaventure, NY: The Franciscan Institute, 2006).

John Scotus Eriugena, *Periphyseon. Liber tertius* (Corpus Christianorum Continuatio Mediaevalis 163), ed. Edouard A. Jeauneau (Turnhout: Brepols, 1999).

Michel de Montaigne, *Les Essais*, ed. Fortunat Strowski (Bordeaux: Pech, 1909).

Nicole Oresme, *Expositio et quaestiones in Aristotelis* De anima, ed. Benoît Patar (Leuven/Paris: Peeters, 1995).

Peter of John Olivi, *Quaestiones in secundum librum Sententiarum*, 3 vols., ed. Bernhard Jansen (Florence: Collegium S. Bonaventurae, 1922–1926).

Physiologus. A Medieval Book of Nature Lore, tr. Michael J. Curley (Chicago/London: University of Chicago Press, 2009).

Peter of Spain, *Questiones super libro 'De Animalibus' Aristotelis. Critical Edition with Introduction*, ed. Francisca Navarro Sánchez (London/New York: Routledge, 2015).

Pseudo-Peter of Spain, *Commentum super libros de animalibus* (Venetian redaction), Venice, Biblioteca Nazionale Marciana, Lat. VI, 234, fol. 1ra-303vb.

Ralph of Battle, *De nesciente et sciente* (*Dialoge zur philosophischen Theologie, Lateinisch-Deutsch*), eds. and tr. Samu Niskanen and Bernd Goebel (Freiburg/Basel/Wien: Herder, 2015).

Roger Bacon, *Communia naturalium I* (Opera hactenus inedita III), ed. Robert Steele (Oxford: Clarendon Press, 1911).

Roger Bacon, *Questiones super quator libros Physicorum* (Opera hactenus inedita VIII), ed. Ferdinand M. Delorme (Oxford: Clarendon Press, 1928).

Roger Bacon, *Questiones supra libros prime philosophie Aristotelis* (*Metaphysica I, II, V-X*) (Opera hactenus inedita X), ed. Robert Steele (Oxford: Clarendon Press, 1930).

Roger Bacon, *Questiones altere supra libros prime philosophie Aristotelis* (*Metaphysica I-IV*) (Opera hactenus inedita XI), ed. Robert Steele (Oxford: Clarendon Press, 1932).

Roger Bacon, *Roger Bacon's Philosophy of Nature. A Critical Edition, with English Translation, Introduction, and Notes, of* De Multiplicatione Specierum *and* De Speculis Comburentibus, ed. David C. Lindberg (Oxford: Clarendon Press, 1983).

Roger Bacon, *Roger Bacon and the Origins of* Perspectiva *in the Middle Ages. A Critical Edition and English Translation of Bacon's* Perspectiva *with Introduction and Notes*, ed. David C. Lindberg (Oxford: Clarendon Press, 1996).

Sextus Empiricus, *Outlines of Pyrrhonism*, 4 vols., tr. Robert G. Bury (London/Cambridge, Mass.: William Heinemann/Harvard University Press, 1933).

Thomas Aquinas, *Commentaria in Aristotelis libros Peri hermeneias et Posteriorum analyticorum*, Editio Leonina, vol. I (Rome 1882).

Thomas Aquinas, *In primum librum Posteriorum analyticorum Aristoteles expositio*, Editio Leonina, vol. I (Rome 1882).

Thomas Aquinas, *Pars prima Summae theologiae, qq. 50–119*, Editio Leonina, vol. V (Rome 1889).

Thomas Aquinas, *Prima secundae Summae theologiae, qq. 1–70*, Editio Leonina, vol. VI (Rome 1891).

Thomas Aquinas, *Secunda secundae Summae theologiae, qq. 1–56*, Editio Leonina, vol. VIII (Rome 1895).

Thomas Aquinas, *Secunda secundae Summae theologiae, qq. 57–122*, Editio Leonina, vol. IX (Rome 1897).

Thomas Aquinas, *Summa contra gentiles, lib. I-II*, Editio Leonina, vol. XIII (Rome 1918).

Thomas Aquinas, *Summa contra gentiles, lib. IV*, Editio Leonina, vol. XV (Rome 1930).

Thomas Aquinas, *Scriptum super Sententiis*, vol. 3, ed. Maria F. Moos (Paris: Lethielleux, 1933).

Thomas Aquinas, *In Aristotelis librum De anima commentarium*, ed. Angelo M. Pirotta (Turin/Rome: Marietti, 1959).

Thomas Aquinas, *In duodecim libros Metaphysicorum Aristotelis expositio*, eds. Marie-Raymond Cathala & Raimondo M. Spiazzi (Turin/Rome: Marietti, 1964).

Thomas Aquinas, *Sententia libri Ethicorum*, Editio Leonina, vol. XLVII.2 (Rome 1969).

Thomas Aquinas, *Quaestiones disputatae De veritate*, Editio Leonina, vol. XXII.3.1 (Rome 1973).

Thomas Aquinas, *Sentencia libri De anima*, Editio Leonina, vol. XLV.1 (Rome 1984).

Thomas Aquinas, *Sentencia libri De sensu et sensato*, Editio Leonina, vol. XLV.2 (Rome 1985).

Thomas Aquinas, *Über die Wahrheit. Teilband 5*, tr. Tilman A. Ramelow (Hamburg: Meiner, 2013).

Thomas of Cantimpré, *Liber de natura rerum*, ed. Helmut Boese (Berlin: Walter de Gruyter, 1973).

William of Ockham, *Quaestiones in librum tertium Sententiarum* (*Reportatio*) (Opera theologica VI), eds. Francis E. Kelley and Girard J. Etzkorn (St. Bonaventure: St. Bonaventure University, 1982).

William of Ockham, *Quaestiones in librum quartum Sententiarum* (*Reportatio*) (Opera theologica VII), eds. Gedeon Gál and Romualdo Green (St. Bonaventure, NY: Franciscan Institute, St. Bonaventure University, 1984).

Secondary Literature

Adams, Marilyn McCord, "Universals in the Early Fourteenth Century," *The Cambridge History of Later Medieval Philosophy. From the Rediscovery of Aristotle to the Disintegration of Scholasticism 1100–1600*, eds. Norman Kretzmann, Anthony Kenny, and Jan Pinborg (Cambridge: Cambridge University Press, 1982), 411–439.

Adamson, Peter (forthcoming), "Human and Animal Nature in the Philosophy of the Islamic World," *Animals*, eds. G. Fay Edwards and Peter Adamson (Oxford: Oxford University Press).

Allen, Colin and Marc D. Hauser (1991), "Concept Attribution in Nonhuman Animals: Theoretical and Methodological Problems in Ascribing Complex Mental Processes," *Philosophy of Science* 58/2, 221–240.

Amerini, Fabrizio (2011), "Intentionality," *Encyclopedia of Medieval Philosophy. Philosophy between 500 and 1500*, vol. 1, ed. Henrik Lagerlund (Springer: Dordrecht et al.), 558–564.

Anzulewicz, Henryk (2002), "Konzeptionen und Perspektiven der Sinneswahrnehmung im System Alberts des Grossen," *Micrologus* X, 199–238.

Anzulewicz, Henryk (2005), "*Memoria* und *reminiscentia* bei Albertus Magnus," *La mémoire du temps au Moyen Age*, ed. Agostino Paravicini Bagliani (Firenze: SISMEL – Edizioni del Galluzzo), 163–200.

Anzulewicz, Henryk (2009), "Albertus Magnus und die Tiere," *Tiere und Fabelwesen im Mittelalter*, ed. Sabine Obermaier (Berlin/New York: Walter de Gruyter), 29–54.

Anzulewicz, Henryk (2013a), "Anthropology: The Concept of Man in Albert the Great," *A Companion to Albert the Great. Theology, Philosophy, and the Sciences*, ed. Irven M. Resnick, (Leiden/Boston: Brill), 325–246.

Anzulewicz, Henryk (2013b), "*Lupus rapax*: Allegorisches und Naturkundliches über den Wolf bei Albertus Magnus," *Reinardus* 25/1, 11–27.

Anzulewicz, Henryk (2016), "Theodor W. Köhler, *Homo animal nobilissimum*," *Theologische Revue* 112/4, 313–318.

Asúa, Miguel de (1991), *The Organization of Discourse on Animals in the Thirteenth Century. Peter of Spain, Albert the Great, and the Commentaries on* 'De animalibus'. PhD Diss. (Ann Arbor: Notre Dame).

Asúa, Miguel de (1994), "El *De Animalibus* de Alberto Magno y la organización del discurso sobre los animales en el siglo XIII," *Patristica et Mediaevalia* 15, 3–26.

Asúa, Miguel de (1997), "Peter of Spain, Albert the Great, and the *Quaestiones de animalibus*," *Physis* 34, 1–30.

Asúa, Miguel de (1999), "Medicine and Philosophy in Peter of Spain's Commentary on *De animalibus*," *Aristotle's Animals in the Middle Ages and Renaissance*, eds.

Carlos G. Steel, Guy Guldentops, and Pieter Beullens (Leuven: Leuven University Press), 189–211.

Bakker, Paul J.J.M., and Johannes M.M.H. Thijssen (eds.) (2007), *Mind, Cognition and Representation. The Tradition of Commentaries on Aristotle's* De Anima (Aldershot: Ashgate).

Balss, Heinrich (1954), "Die Tausendfüßler, Insekten und Spinnen bei Albertus Magnus," *Sudhoffs Archiv für Geschichte der Medizin und der Naturwissenschaften* 38/4, 303–322.

Barad, Judith A. (1995), *Aquinas on the Nature and Treatment of Animals* (San Francisco: International Scholars Publications).

Baranzke, Heike (2002), *Würde der Kreatur? Die Idee der Würde im Horizont der Bioethik* (Würzburg: Königshausen & Neumann).

Barker, Mark J. (2007), *The Cogitative Power. Objects and Terminology*, PhD Diss. (Houston: University of St. Thomas, Center for Thomistic Studies) [unpublished].

Barker, Mark J. (2012a), "Aquinas on Internal Sensory Intentions: Nature and Classification," *International Philosophical Quarterly* 52/2, 199–226.

Barker, Mark J. (2012b), "Experience and Experimentation: The Meaning of *Experimentum* in Aquinas," *The Thomist* 76/3, 37–71.

Barth, Christian (2010), *Objectivity and the Language-Dependence of Thought. A Transcendental Defence of Universal Lingualism* (New York/London: Routledge).

Bennett, Jonathan F. (1989), *Rationality. An Essay Towards an Analysis* (Indianapolis: Hackett).

Bermúdez, José L. (2003), *Thinking without Words* (Oxford/New York: Oxford University Press).

Beullens, Pieter (2007), "Like a Book Written by God's Finger: Animals Showing the Path toward God," *A Cultural History of Animals in the Medieval Age*, ed. Brigitte Resl (Oxford/New York: Berg), 127–151.

Biard, Joël (2012), *Science et nature. La théorie buridanienne du savoir* (Paris: Vrin).

Black, Deborah L. (1993), "Estimation (Wahm) in Avicenna. The Logical and Psychological Dimensions," *Dialogue*, 32/2, 219–258.

Black, Deborah L. (1996), "Memory, Individuals, and the Past in Averroes's Psychology," *Medieval Philosophy and Theology* 5/2, 161–187.

Black, Deborah L. (2000), "Imagination and Estimation: Arabic Paradigms and Western Transformations," *Topoi* 19/1, 59–75.

Black, Deborah L. (2009), "The Nature of Intellect," *The Cambridge History of Medieval Philosophy*, vol. 1, ed. Robert Pasnau (Cambridge: Cambridge University Press), 320–333.

Black, Deborah L. (2011), "Avicenna's 'Vague Individual' and Its Impact on Medieval Latin Philosophy," *Vehicles of Transmission, Translation, and Transformation in Medieval Textual Culture*, eds. Robert Wisnovsky et al. (Turnhout: Brepols), 259–292.

Black, Deborah L. (forthcoming), "Memory, Imagination, Particular Reason, and Thought in Later Medieval Philosophy," *A Companion to Medieval Theories of Cognition*, eds. Russell L. Friedman and Martin Pickavé (Leuven: Leuven University Press).

Bloch, David K. (2007), *Aristotle on Memory and Recollection. Text, Translation, Interpretation, and Reception in Western Scholasticism* (Leiden: Brill).

Bloch, David K. (2014), "Memory and Recollection. Ancient and Medieval Theories," *Sourcebook for the History of the Philosophy of Mind. Philosophical Psychology from Plato to Kant*, eds. Simo Knuuttila and Juha Sihvola (Dordrecht: Springer), 205–221.

Brandom, Robert (2000), *Articulating Reasons. An Introduction to Inferentialism* (Cambridge, Mass./London: Harvard University Press).

Brittain, Charles (2002), "Non-Rational Perception in the Stoics and Augustine," *Oxford Studies in Ancient Philosophy* XXII, 253–308.

Brungs, Alexander (2002), *Metaphysik der Sinnlichkeit. Das System der Passiones Animae bei Thomas von Aquin* (Halle: Hallescher Verlag).

Buchenau, Stefanie, and Roberto Lo Presti (eds.) (2017), *Human and Animal Cognition in Early Modern Philosophy and Medicine* (Pittsburgh: University of Pittsburgh Press).

Burnett, Charles (ed.) (1987), *Adelard of Bath. An English Scientist and Arabist of the Early Twelfth Century* (London: Warburg Institute).

Call, Josep, and Michael Tomasello (2008), "Does the Chimpanzee Have A Theory of Mind? 30 Years Later," *Trends in Cognitive Sciences* 12/5, 187–192.

Carruthers, Mary J. (1992), *The Book of Memory. A Study of Memory in Medieval Culture* (Cambridge: Cambridge University Press).

Cassin, Barbara, and Jean-Louis Labarrière (eds.) (1997), *L'animal dans l'Antiquité* (Paris: Vrin).

Cheke, Lucy G., and Nicola S. Clayton (2010), "Mental Time Travel in Animals," *Wiley Interdisciplinary Reviews. Cognitive Science* 1/6, 915–930.

Chenu, Marie-Dominique (1968), "Nature and Man. The Renaissance of the Twelfth Century," *Nature, Man, and Society in the Twelfth Century. Essays on New Theological Perspectives in the Latin West*, eds. Jerome Taylor and Lester K. Little (Chicago/London: University of Chicago Press), 1–48.

Cheung, Tobias (ed.) (2010), *Transitions and Borders between Animals, Humans and Machines 1600–1800* (Leiden: Brill).

Clark, Gillian (1998), "The Fathers and the Animals. The Rule of Reason?," *Animals on the Agenda. Questions about Animals for Theology and Ethics*, eds. Andrew Linzey and Dorothy Yamamoto (London: SCM Press), 67–79.

Clark, Willene B. (1995), "Zoology in the Medieval Latin Bestiary," *Man and Nature in the Middle Ages*, eds. Susan J. Ridyard and Robert B. Benson (Sewanee: University of the South), 223–245.

Clark, Willene B. (2006), *A Medieval Book of Beasts. The Second-Family Bestiary: Commentary, Art, Text and Translation* (Woodbridge: Boydell Press).

Clark, Willene B., and Meredith McCunn (eds.) (1989), *Beasts and Birds of the Middle Ages. The Bestiary and its Legacy* (Philadelphia: University of Pennsylvania Press).

Clarke, Edwin, and Kenneth Dewhurst (1996), *An Illustrated History of Brain Function. Imaging the Brain from Antiquity to the Present*, 2nd ed. (San Francisco: Norman Publishing).

Cohen, Jeremy (1992), *'Be Fertile and Increase, Fill the Earth and Master It'. The Ancient and Medieval Career of a Biblical Text* (Ithaca/London: Cornell University Press).

Coleman, Janet (1992), *Ancient and Medieval Memories. Studies in the Reconstruction of the Past* (Cambridge: Cambridge University Press).

Coles, Andrew (1997), "Animal and Childhood Cognition in Aristotle's Biology and the *scala naturae*," *Aristotelische Biologie. Intentionen, Methoden, Ergebnisse*, eds. Wolfgang Kullmann and Sabine Föllinger (Stuttgart: Steiner), 287–323.

Cottingham, John (1978), "'A Brute to the Brutes?': Descartes' Treatment of Animals," *Philosophy* 53/206, 551–559.

Cova, Luciano (1997), "Il Corpus zoologico di Aristotele nei dibatti fra gli 'artisti' parigini alle soglie del XIV secolo," *L'enseignement des disciplines à la Faculté des arts*, eds. Olga Weijers and Louis Holtz (Turnhout: Brepols), 281–302.

Crane, Tim (1998), "Intentionality," *Routledge Encyclopedia of Philosophy*, vol. 4, ed. Edward Craig (London: Routledge), 816–821.

Crisciani, Chiara, Roberto Lambertini, and Romana Martorelli (eds.) (2004), *Parva naturalia. Saperi medievali, natura e vita* (Pisa: Istituti editoriali e poligrafici internazionali).

D'Ancona, Cristina (2008), "Degrees of Abstraction in Avicenna," *Theories of Perception in Medieval and Early Modern Philosophy*, eds. Simo Knuuttila and Pekka Kärkkäinen (Dordrecht: Springer), 47–72.

Daston, Lorraine (2005), "Intelligences: Angelic, Animal, Human," *Thinking with Animals. New Perspectives on Anthropomorphism*, eds. Gregg Mitman and id. (New York: Columbia University Press), 37–58.

Daston, Lorraine, and Gregg Mitman (eds.) (2005), *Thinking with Animals. New Perspectives on Anthropomorphism* (New York: Columbia University Press).

Davids, Tobias (2017), *Anthropologische Differenz und animalische Konvenienz. Tierphilosophie bei Thomas von Aquin* (Leiden/Boston: Brill).

Davidson Donald (1982), "Rational Animals," *Dialectica* 46/4, 317–327.

De Boer, Sander W. (2013), *The Science of the Soul. The Commentary Tradition on Aristotle's* De anima, *c. 1260–c. 1360* (Leuven: Leuven University Press).

De Boer, Sander W. (2014), "John Buridan on the Internal Senses," *Documenti e Studi sulla Tradizione Filosofica Medievale* XXV, 403–421.

Deeley, John N. (1971), "Animal Intelligence and Concept-Formation," *The Thomist* 35/1, 43–93.

De Haan, Daniel D. (2014), "Perception and the *Vis Cogitativa*. A Thomistic Analysis of Aspectual, Actional, and Affectional Percepts," *American Catholic Philosophical Quarterly* 88/3, 397–437.

De Leemans, Pieter (1999), "The Vicissitudes of a Zoological Treatise: Aristotle's *De incessu animalium* in the Middle Ages and Renaissance," *Tradition et traduction. Les textes philosophiques et scientifiques grec au moyen âge latin*, eds. Rita Beyers et al. (Leuven: Leuven University Press), 199–218.

De Leemans, Pieter (2000), "Medieval Latin Commentaries on Aristotle's '*De motu animalium*'," *Recherches de Théologie et Philosophie médiévales* 67/2, 272–360.

De Leemans, Pieter (2005), "Internal Senses, Intellect and Movement. Peter of Auvergne (?) on Aristotle's De Motu Animalium," *Corpo e anima, sensi interni e intelletto dai secoli XIII-XIV ai post-cartesiani e spinoziani*, ed. Graziella Federici Vescovini, Valeria Sorge, and Carlo Vinti (Turnhout: Brepols), 139–160.

De Leemans, Pieter, and Matthew Klemm (2007), "Animals and Anthropology in Medieval Philosophy," *A Cultural History of Animals in the Medieval Age*, ed. Brigitte Resl (Oxford/New York: Berg), 153–177.

Dennett, Daniel C. (1971), "Intentional Systems," *The Journal of Philosophy* 68/4, 87–106.

De Waal, Frans (1999), "Anthropomorphism and Anthropodenial: Consistency in Our Thinking about Humans and Other Animals," *Philosophical Topics* 27/1, 255–280.

Dewender, Thomas (2003), "Zur Rezeption der Aristotelischen Phantasialehre in der lateinischen Philosophie des Mittelalters," *Imagination - Fiktion - Kreation. Das kulturschaffende Vermögen der Phantasie*, eds. Thomas Dewender and Thomas Welt (München/Leipzig: K.G. Saur), 141–160.

Dickerman, Sherwood O. (1911), "Some Stock Illustrations of Animal Intelligence in Greek Psychology," *Transactions and Proceedings of the American Philological Association* 42, 123–130.

Dierauer, Urs (1977), *Tier und Mensch im Denken der Antike. Studien zur Tierpsychologie, Anthropologie und Ethik* (Amsterdam: Grüner).

Di Martino, Carla (2006), "*Memoria dicitur multipliciter*. L'apporto della scienza psicologica araba al medievo latino," *Tracce nella mente. Teorie della memoria da Platone ai moderni*, ed. Maria M. Sassi (Pisa: Edizioni della Scuola normale superiore), 119–138.

Di Martino, Carla (2007), "Memory and Recollection in Ibn Sînâ's and Ibn Rushd's Philosophical Texts Translated into Latin in the Twelfth and Thirteenth Centuries: A Perspective on the Doctrine of the Internal Senses in Arabic Psychological Science," *Forming the Mind. Essays on the Internal Senses and the Mind/Body Problem from Avicenna to the Medical Enlightenment*, ed. Henrik Lagerlund (Dordrecht: Springer), 17–26.

Di Martino, Carla (2008), *Ratio particularis. La doctrine des sens internes d'Avicenne à Thomas d'Aquin. Contribution à l'étude de la tradition arabo-latine de la psychologie d'Aristote* (Paris: Vrin).

Dinzelbacher, Peter (2000), "Mittelalter," *Mensch und Tier in der Geschichte Europas*, ed. id. (Stuttgart: Kröner), 181–292.

Dod, Bernard G. (1982), "Aristoteles Latinus," *The Cambridge History of Later Medieval Philosophy. From the Rediscovery of Aristotle to the Disintegration of Scholasticism 1100–1600*, eds. Norman Kretzmann, Anthony Kenny, and Jan Pinborg (Cambridge: Cambridge University Press), 43–79.

Dohm, Burkhard (2013), "Die Unsterblichkeit der Tierseele im Denken des Nicolaus Cusanus," *The Animal Soul and the Human Mind. Renaissance Debates*, ed. Cecilia Muratori (Pisa/Rome: Serra), 27–44.

Dretske, Fred (1988), *Explaining Behavior. Reasons in a World of Causes* (Cambridge, Mass./London: MIT Press).

Dretske, Fred (2006), "Minimal Rationality," *Rational Animals?*, eds. Susan L. Hurley and Matthew Nudds (Oxford: Oxford University Press), 107–115.

Dronke, Peter (1985), "La creazione degli animali," *L'uomo di fronte al mondo animale nell'alto medioevo*, vol. 2 (Spoleto: Presso la Sede del Centro), 809–842.

Drossaart-Lulofs, Hendrik J. (1992), "Preface," *Aristotle, De animalibus. Michael Scot's Arabic-Latin Translation. Part Three. Books XV-XIX: Generation of Animals*, ed. Aafke M.I. Van Oppenraaij (Leiden/New York/Köln: Brill), vii–xvi.

Druart, Thérèse-Anne (1980), "Le traité d'Avempace sur 'Les choses au moyen desquelles on peut connaître l'intellect agent'," *Bulletin de Philosophie Médiévale* 22, 73–77.

Druart, Thérèse-Anne (2016), "Logic and Language," *The Routledge Companion to Islamic Philosophy*, eds. Richard C. Taylor and Luis Xavier López-Farjeat (New York/Oxford: Routledge), 69–81.

Eco, Umberto, Costantino Marmo, Roberto Lambertini, and Andrea Tabarroni (1989), "On Animal Language in the Medieval Classification of Signs," *On the Medieval Theory of Signs*, eds. Umberto Eco and Constantine Marmo (Amsterdam/Philadelphia: John Benjamins), 3–32.

Enders, Markus (2001), "Vernunft; Verstand (III. Mittelalter – A. Augustinus, Frühmittelalter, Frühscholastik)," *Historisches Wörterbuch der Philosophie*, vol. 11 (Darmstadt: Wissenschaftliche Buchgesellschaft), 764–770.

Federici Vescovini, Graziella (2004), "La tradizione dei *Parva naturalia* nell' insegnamento universitario medievale (secoli XIII e XIV)," *Parva naturalia. Saperi medievali, natura e vita*, eds. Crisciani Chiara, Roberto Lambertini, and Romana Martorelli (Pisa: Istituti editoriali e poligrafici internazionali), 125–141.

Federici Vescovini, Graziella, Valeria Sorge, and Carlo Vinti (eds.) (2005), *Corpo e anima, sensi interni e intelletto dai secoli XIII-XIV ai post-cartesiani e spinoziani* (Turnhout: Brepols).

Flores, Nora C. (ed.) (1996), *Animals in the Middle Ages. A Book of Essays* (New York/London: Garland Publishing).

Floridi, Luciano (1997), "Scepticism and Animal Rationality: the Fortune of Chrysippus' Dog in the History of Western Thought," *Archiv für Geschichte der Philosophie* 79/1, 27–57.

Frampton, Michael (2008), *Embodiments of Will. Anatomical and Physiological Theories of Voluntary Animal Motion from Greek Antiquity to the Latin Middle Ages, 400 B.C.-A.D. 1300* (Saarbrücken: VDM Verlag Dr. Müller).

Frede, Dorothea (2001), "Aquinas on *Phantasia*," *Ancient and Medieval Theories of Intentionality*, ed. Dominik Perler (Leiden/Boston/Köln: Brill), 155–183.

Friedrich, Udo (2009), *Menschentier und Tiermensch. Diskurse der Grenzziehung und Grenzüberschreitung im Mittelalter* (Göttingen: Vandenhoeck & Ruprecht).

Friedrich, Udo (2015), "Die Paradigmatik des Esels im enzyklopädischen Schrifttum des Mittelalters und der frühen Neuzeit," *Zeitschrift für Germanistik* XXV/1, 93–109.

Fudge, Erica (2006), *Brutal Reasoning. Animals, Rationality, and Humanity in Early Modern England* (Ithaca: Cornell University Press).

Gätje, Helmut (1988), "Gedächtnis und Erinnerung bei Avicenna und Averroes," *Acta Orientalia* 49, 7–36.

Garceau, Benoit (1968), *Iudicium. Vocabulaire, sources, doctrine de Saint Thomas d'Aquin* (Montreal/Paris: Institut d'Études Médiévales/Vrin).

George, Wilma, and Brunsdon Yapp (1991), *The Naming of the Beasts. Natural History in the Medieval Bestiary* (London: Duckworth).

Gilhus, Ingvild Sælid (2006), *Animals, Gods and Humans. Changing Attitudes to Animals in Greek, Roman, and Early Christian Ideas* (London/New York: Routledge).

Glock, Hans-Johann (1999), "Animal Minds: Conceptual Problems," *Evolution and Cognition* 5, 174–188.

Glock, Hans-Johann (2010), "Can Animals Judge?," *Dialectica* 64/1, 11–33.

Glock, Hans-Johann (2012), "The Anthropological Difference: What Can Philosophers Do to Identify the Differences between Human and Non-Human Animals?," *Royal Institute of Philosophy Supplement* 70, 105–131.

Glock, Hans-Johann (2013), "Mental Capacities and Animal Ethics," *Animal Minds & Animal Ethics. Connecting Two Separate Fields*, eds. Klaus Petrus and Markus Wild (Berlin/Bielefeld: Transcript), 113–146.

Glock, Hans-Johann (2016a), "Geist der Tiere," *Tiere. Kulturwissenschaftliches Handbuch*, ed. Roland Borgards (Stuttgart: Metzler), 60–78.

Glock, Hans-Johann (2016b), "Wann ist ein Tier ein Tier? Zum Tierbegriff in Philosophie und Biologie," *Tierisch! Das Tier und die Wissenschaft. Ein Streifzug durch die Disziplinen*, eds. Meret Fehlmann, Margot Michel, and Rebecca Niederhauser (Zürich: vdf Hochschulverlag AG), 13–22.

Granger, Herbert (1985), "The Scala Naturae and the Continuity of Kinds," *Phronesis* 30/2, 181–200.

Grant, Edward (2007), *A History of Natural Philosophy. From the Ancient World to the Nineteenth Century* (Cambridge: Cambridge University Press).

Grimm, Herwig, and Markus Wild (2016), *Tierethik zur Einführung* (Hamburg: Junius).

Guldentops, Guy (1999), "The Sagacity of the Bees. An Aristotelian Topos in Thirteenth-Century Philosophy," *Aristotle's Animals in the Middle Ages and Renaissance*, eds. Carlos G. Steel, Guy Guldentops, and Pieter Beullens (Leuven: Leuven University Press), 275–296.

Guldentops, Guy (2000), "Albert the Great's Zoological Anthropocentrism," *Micrologus* VIII/1, 217–235.

Hackett, Jeremiah (1997), "The Published Works of Roger Bacon," *Vivarium* 35/2, 315–320.

Hackett, Jeremiah (2008), "Roger Bacon's Concept of Experience: A New Beginning in Medieval Philosophy?," *The Modern Schoolman* 86/1, 123–146.

Hackett, Jeremiah (2013), "Animal and Human Knowledge in the *Perspectiva*: (*Opus maius*, Part Five) of Roger Bacon," *Philosophical Psychology in Arabic Thought and the Latin Aristotelianism of the 13th Century*, eds. Luis Xavier López-Farjeat and Jörg Alejandro Tellkamp (Paris: Vrin), 222–241.

Hall, Robert E. (2006), "The *Wahm* in Ibn Sina's Psychology," *Intellect et imagination dans la Philosophie Médiévale*, vol. 1, eds. Maria C. Pacheco and José F. Meirinhos (Turnhout: Brepols), 533–549.

Hamesse, Jacqueline (1988), "'*Imaginatio*' et '*phantasia*' chez les auteurs philosophiques du 12e et du 13e siècle," *Phantasia-Imaginatio*, eds. Marta Fattori and Massimo L. Bianchi (Roma: Ed. dell'Ateneo), 157–165.

Hanus, Daniel et al. (2011), "Comparing the Performances of Apes (*Gorilla gorilla, Pan troglodytes, Pongo pygmaeus*) and Human Children (*Homo sapiens*) in the Floating Peanut Task', *PLoS ONE* 6/6, 1–13.

Harden, Alastair (2013), *Animals in the Classical World. Ethical Perspectives from Greek and Roman Texts* (Basingstoke: Palgrave Macmillan).

Harvey, E. Ruth (1975), *The Inward Wits. Psychological Theory in the Middle Ages and the Renaissance* (London: Warburg Institute, University of London).

Hasse, Dag Nikolaus (2000), *Avicenna's* De anima *in the Latin West. The Formation of a Peripatetic Philosophy of the Soul 1160–1300* (London/Turin: The Warburg Institute – Nino Aragno Editore).

Hasse, Dag Nikolaus (2001), "Avicenna on Abstraction," *Aspects of Avicenna*, ed. Robert Wisnovsky (Princeton: Markus Wiener), 39–72.

Hasse, Dag Nikolaus (2010), "The Soul's Faculties," *The Cambridge History of Medieval Philosophy*, vol. 1, ed. Robert Pasnau (Cambridge: Cambridge University Press), 305–319.

Hassig, Debra (1995), *Medieval Bestiaries. Text, Image, Ideology* (Cambridge/New York: Cambridge University Press).

Heller-Roazen, Daniel (2007), *The Inner Touch. Archaeology of a Sensation* (New York: Zone Books).

Heller-Roazen, Daniel (2008), "Common Sense. Greek, Arabic, Latin," *Rethinking the Medieval Senses. Heritage/Fascinations/Frames*, eds. Stephen G. Nichols, Andreas Kablitz, and Alison Calhoun (Baltimore: The Johns Hopkins University Press), 30–50.

Hicks, Carola (1993), *Animals in Early Medieval Art* (Edinburgh: Edinburgh University Press).

Hoenen, Maarten J.F.M. (1993), "Die Intellektlehre des Johannes Buridan – ihre Quellen und historisch-doktrinären Bezüge," *John Buridan. A Master of Arts. Some Aspects of his Philosophy*, ed. Egbert P. Bos (Nijmegen: Ingenium Publishers), 89–106.

Hoenen, Maarten J.F.M. (2001), "Vernunft; Verstand (III. Mittelalter – F. Spätmittelalterliche Schulen)," *Historisches Wörterbuch der Philosophie*, vol. 11 (Darmstadt: Wissenschaftliche Buchgesellschaft), 790–793.

Hoffman, Paul (2012), "Reasons, Causes, and Inclinations," *Emotion and Cognitive Life in Medieval and Early Modern Philosophy*, eds. Martin Pickavé and Lisa Shapiro (Oxford: Oxford University Press), 156–175.

Holopainen, Toivo J. (2014), "Concepts and Concept Formation in Medieval Philosophy," *Sourcebook for the History of the Philosophy of Mind. Philosophical Psychology from Plato to Kant*, eds. Simo Knuuttila and Juha Sihvola (Dordrecht: Springer), 263–279.

Hünemörder, Christian (1980), "Die Zoologie des Albertus Magnus," *Albertus Magnus. Doctor universalis 1280–1980*, eds. Gerbert Meyer and Albert Zimmermann (Mainz: Matthias-Grünewald-Verlag), 235–248.

Hünemörder, Christian (1988), "Thomas von Aquin und die Tiere," *Thomas von Aquin. Werk und Wirkung im Licht neuerer Forschungen*, eds. Albert Zimmermann and Clemens Kopp (Berlin/New York: Walter de Gruyter), 379–392.

Hünemörder, Christian (2003), "Traditionelle Naturkunde, realistische Naturbeobachtung und theologische Naturdeutung in Enzyklopädien des Hohen Mittelalters," *Natur im Mittelalter. Konzeptionen – Erfahrungen – Wirkungen*, ed. Peter Dilg (Berlin: Akademie Verlag), 124–135.

Hurley, Susan L., and Matthew Nudds (eds.) (2006a), *Rational Animals?* (Oxford: Oxford University Press).

Hurley, Susan L., and Matthew Nudds (2006b), "The Questions of Animal Rationality: Theory and Evidence," *Rational Animals?*, eds. id. (Oxford: Oxford University Press), 1–83.

Ingensiep, Hans Werner (2001), *Geschichte der Pflanzenseele. Philosophische und biologische Entwürfe von der Antike bis zur Gegenwart* (Stuttgart: Kröner).

Iribarren, Isabel, and Martin Lenz (eds.) (2016), *Angels in Medieval Philosophical Inquiry. Their Function and Significance* (London/New York: Routledge).

Janson, Horst W. (1952), *Apes and Ape Lore in the Middle Ages and the Renaissance* (London: Warburg Institute, University of London).

Jones, Richard (2013), *The Medieval Natural World* (London/New York: Routledge).

Juanola, Joan D. A. (2015), "Inteligencia animal y *vis aestimativa* en Avicena y Tomás de Aquino," *Espíritu. Cuadernos del Instituto Filosófico de Balmesiana* 64/150, 341–362.

Kacelnik, Alex, "Meanings of Rationality," *Rational Animals?*, eds. Susan L. Hurley and Matthew Nudds (Oxford: Oxford University Press), 87–106.

Kärkkäinen, Pekka (2011), "Internal Senses," *Encyclopedia of Medieval Philosophy*, vol. 1, ed. Henrik Lagerlund (Dordrecht: Springer), 564–567.

Kann, Christoph (2003), "Zeichen – Ordnung – Gesetz: Zum Naturverständnis in der mittelalterlichen Philosophie," *Natur im Mittelalter. Konzeptionen – Erfahrungen – Wirkungen*, ed. Peter Dilg (Berlin: Akademie Verlag), 34–49.

Kaukua, Jari (2014), "Avicenna on the Soul's Activity in Perception," *Active Perception in the History of Philosophy*, eds. José Filipe Silva and Mikko Yrjönsuuri (Cham et al.: Springer), 99–116.

Kemp, Simon, and Garth J. O. Fletcher (1993), "The Medieval Theory of the Inner Senses," *The American Journal of Psychology* 106/4, 559–576.

Kenny, Anthony (1993), *Aquinas on Mind* (London/New York: Routledge).

King, Peter (2001), "John Buridan's Solution to the Problem of Universals," *The Metaphysics and Natural Philosophy of John Buridan*, eds. Johannes M.M.H. Thijssen and Jack Zupko (Leiden/Boston/Köln: Brill), 1–27.

King, Peter (2008), "The Inner Cathedral: Mental Architecture in High Scholasticism," *Vivarium* 46/3, 253–274.

Kiser, Lisa J. (2007), "Animals in Medieval Sports, Entertainment, and Menageries," *A Cultural History of Animals in the Medieval Age*, ed. Brigitte Resl (Oxford/New York: Berg), 103–126.

Klein, Martin (2016), *Immaterialität, Materialität, Intentionalität. Johannes Buridans Philosophie des Geistes*, PhD Diss. (Berlin: Humboldt-Universität zu Berlin) [unpublished].

Klima, Gyula (2015), "Mental Representations and Concepts in Medieval Philosophy," *Intentionality, Cognition, and Mental Representation in Medieval Philosophy*, ed. id. (New York: Fordham University Press), 323–337.

Klingender, Francis (1971), *Animals in Art and Thought to the End of the Middle Ages* (Cambridge, Mass: MIT Press).

Klubertanz, George P. (1952), *The Discursive Power. Sources and Doctrine of the* Vis Cogitativa *According to St. Thomas Aquinas* (St. Louis: The Modern Schoolman).

Knuuttila, Simo (2008), "Aristotle's Theory of Perception and Medieval Aristotelianism," *Theories of Perception in Medieval and Early Modern Philosophy*, eds. Simo Knuuttila and Pekka Kärkkäinen (Dordrecht: Springer), 1–22.

Knuuttila, Simo, and Pekka Kärkkäinen (2014), "Medieval Theories of Internal Senses," *Sourcebook for the History of the Philosophy of Mind. Philosophical Psychology from Plato to Kant*, eds. Simo Knuuttila and Juha Sihvola (Dordrecht: Springer), 131–145.

Koch, Joseph (1931), "Sind die Pygmäen Menschen? Ein Kapitel aus der philosophischen Anthropologie der mittelalterlichen Scholastik," *Archiv für Geschichte der Philosophie* 40/2, 194–213.

Köhler, Theodor W. (1992), "Anthropologische Erkennungsmerkmale menschlichen Seins. Die Frage der ‚Pygmei' in der Hochscholastik," *Mensch und Natur im Mittelalter*, eds. Albert Zimmermann and Andreas Speer (Berlin/New York: Walter de Gruyter), 718–735.

Köhler, Theodor W. (1999), "Die wissenschaftstheoretische und inhaltliche Bedeutung der Rezeption von *De animalibus* für den philosophisch-anthropologischen Diskurs im 13. Jahrhundert," *Aristotle's Animals in the Middle Ages and Renaissance*, eds. Carlos G. Steel, Guy Guldentops, and Pieter Beullens (Leuven: Leuven University Press), 249–274.

Köhler, Theodor W. (2000), *Grundlagen des philosophisch-anthropologischen Diskurses im dreizehnten Jahrhundert. Die Erkenntnisbemühung um den Menschen im zeitgenössischen Verständnis* (Leiden/Boston/Köln: Brill).

Köhler, Theodor W. (2001), "Der Tiervergleich als philosophisch-anthropologisches Schlüsselparadigma – der Beitrag Alberts des Großen," *Albertus Magnus. Zum Gedenken nach 800 Jahren: Neue Zugänge, Aspekte und Perspektiven*, ed. Walter Senner (Berlin: Akademie Verlag), 437–454.

Köhler, Theodor W. (2006), "*Imaginatio/phantasia* bei Mensch und Tier. Aspekte philosophischer Urteilsbildung nach veröffentlichten und unveröffentlichten Quellen des 13. Jahrhunderts," *Intellect et imagination dans la Philosophie Médiévale, vol. 2*, eds. Maria C. Pacheco and José F. Meirinhos (Turnhout: Brepols), 1085–1098.

Köhler, Theodor W. (2008), *Homo animal nobilissimum. Konturen des spezifisch Menschlichen in der naturphilosophischen Aristoteleskommentierung des dreizehnten Jahrhunderts, Teilband 1* (Leiden/Boston: Brill).

Köhler, Theodor W. (2014), *Homo animal nobilissimum. Konturen des spezifisch Menschlichen in der naturphilosophischen Aristoteleskommentierung des dreizehnten Jahrhunderts, Teilband 2.1 und 2.2* (Leiden/Boston: Brill).

König-Pralong, Catherine (2011), "Animal equivoque: De Lincoln à Paris via Cologne," *Mots médiévaux offerts à Ruedi Imbach*, ed. Iñigo Atucha (Porto: Fédération Internationale des Instituts d'Études Médiévales), 67–76.

Kruk, Remke (1995), "Traditional Islamic Views of Apes and Monkeys," *Ape, Man, Apeman. Changing Views Since 1600*, eds. Raymond Corbey and Bert Theunissen (Leiden: Department of Prehistory, Leiden University), 29–42.

Kruk, Remke (1997), "Ibn Bajja's Commentary on Aristotle's *De animalibus*," *The Ancient Tradition in Christian and Islamic Hellenism*, eds. Remke Kruk and Gerhard Endress (Leiden: CNWS), 165–180.

Kruk, Remke (2001), "Timotheus of Gaza's *On Animals* in the Arabic Tradition," *Le Muséon* 114/3, 355–387.

Kruk, Remke (2002), "Ibn Sīnā *On Animals*: Between the First Teacher and the Physician," *Avicenna and His Heritage*, eds. Jules L. Janssens and Daniel de Smet (Leuven: Leuven University Press), 325–341.

Kruk, Remke (2003), "La zoologie aristotélicienne. Tradition arabe," *Dictionnaire des philosophes antiques. Supplément*, ed. Richard Goulet (Paris: CNRS), 329–334.

Labarrière, Jean-Louis (2004), *Langage, vie politique et mouvement des animaux* (Paris: Vrin).

Lagerlund, Henrik (2017), "Awareness and Unity of Conscious Experience: Buridan on the Common Sense," *Questions on the Soul by John Buridan and Others. A Companion to John Buridan's Philosophy of Mind*, ed. Guyla Klima (Cham: Springer), 149–156.

Lauzi, Egle (2012), *Il destino degli animali. Aspetti delle tradizioni culturali araba e occidentale nel Medio Evo* (Firenze: SISMEL – Edizioni del Galluzzo).

Li Causi, Pietro, and Roberto Pomelli (eds.) (2015), *L'anima degli animali. Aristotele, frammenti stoici, Plutarco, Porfirio* (Torino: Einaudi).

Lindberg, David C. (1976), *Theories of Vision from Al-Kindi to Kepler* (Chicago: University of Chicago Press).

Lisska, Anthony J. (2006), "A Look at Inner Sense in Aquinas: A Long-Neglected Faculty Psychology," *Proceedings of the American Catholic Philosophical Association* 80, 1–19.

Lisska, Anthony J. (2016), *Aquinas's Theory of Perception. An Analytic Reconstruction* (Oxford: Oxford University Press).

Lohr, Charles H. (1982), "The Medieval Interpretation of Aristotle," *The Cambridge History of Later Medieval Philosophy. From the Rediscovery of Aristotle to the Disintegration of Scholasticism 1100–1600*, eds. Norman Kretzmann, Anthony Kenny, and Jan Pinborg (Cambridge: Cambridge University Press), 80–98.

López-Farjeat, Luis Xavier (2010), "El conocimiento animal en Aristóteles y Avicena," *Acta Philosophica* 20, 125–144.

López-Farjeat, Luis Xavier (2012), "Self-Awareness (*Al-Shu'Ūr Bi-Al-Dhāt*) in Human and Non-Human Animals in Avicenna's Psychological Writings," *Oikeiosis and the Natural Bases of Morality. From Classical Stoicism to Modern Philosophy*, ed. Alejandro G. Vigo (Hildesheim/Zürich/New York: Olms), 121–140.

López-Farjeat, Luis Xavier (2013), "¿Tienen los animales no humanos un yo? Una posible respuesta desde la filosofía de la mente de Avicena," *Signos Filosóficos* 15/30, 71–88.

López-Farjeat, Luis Xavier (2016), "Avicenna on Non-Conceptual Content and Self-Awareness in Non-Human Animals," *Subjectivity and Selfhood in Medieval and Early Modern Philosophy*, eds. Jari Kaukua and Tomas Ekenberg (Dordrecht: Springer), 61–73.

Loughlin, Stephen (2001), "Similarities and Differences between Human and Animal Emotion in Aquinas's Thought," *The Thomist* 65, 45–65.

Lovejoy, Arthur O. (1936), *The Great Chain of Being. A Study of the History of an Idea* (Cambridge, Mass./London: Harvard University Press).

Lurz, Robert W. (ed.) (2009), *The Philosophy of Animal Minds* (Cambridge: Cambridge University Press).

MacIntyre, Alasdair C. (1999), *Dependent Rational Animals. Why Human Beings Need the Virtues* (Chicago: Duckworth).

Maierù, Alfonso (2011), "Individuum vagum. L'influenza di Avicenna in un commento a Porfirio della prima metà del secolo XIII," *Mots médiévaux offerts à Ruedi Imbach*, eds. Iñigo Atucha et al. (Porto: Fédération Internationale des Instituts d'Études Médiévales), 353–364.

Malcolm, Norman (1972), "Thoughtless Brutes," *Proceedings and Addresses of the American Philosophical Association* 46, 5–20.

Maloney, Thomas S. (1985), "The Extreme Realism of Roger Bacon," *The Review of Metaphysics* 38/4, 807–837.

Manzanedo, Marcos F. (1989), "¿Poseen inteligencia los animales?," *Ciencia tomista* 116, 99–128.

Manzanedo, Marcos F. (1990), "La cogitativa del hombre y la inteligencia de los animales," *Angelicum* 67 /3, 329–363.

Matthews, Gareth B. (1978), "Animals and the Unity of Psychology," *Philosophy* 53/206, 437–454.

Matthews, Gareth B. (1999), "Augustine and Descartes on the Souls of Animals," *From Soul to Self*, ed. M. James C. Crabbe (London: Routledge), 89–107.

Menzel, Randolf, and Julia Fischer (2011), "Animal Thinking. An Introduction," *Animal Thinking. Contemporary Issues in Comparative Cognition*, eds. id. (Cambridge, Mass.: MIT Press), 1–6.

Meo-Ehlert, Myrta de (ed.) (2015), *Confini animali dell'anima umana. Prospettive e problematiche* [= special issue of *Lo Sguardo* 18/2].

Michon, Cyrille (2001), "Intentionality and Proto-Thoughts," *Ancient and Medieval Theories of Intentionality*, ed. Dominik Perler (Leiden/Boston/Köln: Brill), 325–341.

Miner, Robert (2009), *Thomas Aquinas on the Passions. A Study of* Summa Theologiae *1a2ae 22–48* (Cambridge: Cambridge University Press).

Moretti, Paola F. (2005), "Cane d'un filosofo, cane d'un eretico. Appunti sulla fortuna cristiana del cane sillogistico," *Nuovo e antico nella cultura greco-latina di IV-VI secolo*, eds. Isabella Gualandri, Fabrizio Conca, and Raffaele Passarella (Milano: Cisalpino), 633–660.

Morgan, Conwy Lloyd (1894), *An Introduction to Comparative Psychology* (London: Walter Scott).

Müller, Jörn (2015a), "Memory as an Internal Sense. Avicenna and the Reception of His Psychology by Thomas Aquinas," *Quaestio* 15, 45–54.

Müller, Jörn (2015b), "Memory in Medieval Philosophy," *Memory. A History*, ed. Dmitri Nikulin (New York: Oxford University Press), 92–124.

Müller, Jörn (2017), *Albertus Magnus über Gedächtnis, Erinnern und Wiedererinnerung. Eine philosophische Lektüre von* De memoria et reminiscentia *mit Übersetzung* (Münster: Aschendorff).

Muratori, Cecilia (ed.) (2013), *The Animal Soul and the Human Mind. Renaissance Debates* (Pisa/Rome: Serra).

Muratori, Cecilia, and Burkhard Dohm (eds.) (2013), *Ethical Perspectives on Animals in the Renaissance and Early Modern Period* (Firenze: SISMEL – Edizioni del Galluzzo).

Newmyer, Stephen T. (2011), *Animals in Greek and Roman Thought. A Sourcebook* (Abingdon/New York: Routledge).

Nitschke, August (1967), "Verhalten und Bewegung der Tiere nach frühen christlichen Lehren," *Studium Generale* 4/20, 235–262.

Nuchelmans, Gabriël (1973), *Theories of the Proposition. Ancient and Medieval Conceptions of the Bearers of Truth and Falsity* (Amsterdam/London: North-Holland Publishing Company).

Nuchelmans, Gabriël (1980), *Late-Scholastic and Humanist Theories of the Proposition* (Amsterdam/New York: North-Holland Publishing Company).

Nussbaum, Martha C. (1985), "The Role of *Phantasia* in Aristotle's Explanations of Action," *Aristotle's De motu animalium. Text with Translation, Commentary, and Interpretive Essays*, ed. id. (Princeton: Princeton University Press), 221–269.

Obermaier, Sabine (2009), "Tiere und Fabelwesen im Mittelalter. Einführung und Überblick," *Tiere und Fabelwesen im Mittelalter*, ed. id. (Berlin/New York: Walter de Gruyter), 1–25.

Oelze, Anselm (forthcoming a), "Geschichte der Tierethik: Mittelalter," *Handbuch Tierethik. Grundlagen – Kontexte - Perspektiven*, eds. Johann S. Ach and Dagmar Borchers (Stuttgart: Metzler).

Oelze, Anselm (forthcoming b), "Können Tiere irren? Philosophische Antworten aus dem 13. und 14. Jahrhundert," *Irrtum. Error. Erreur*, eds. Andreas Speer and Maxime Mauriège, (Berlin/Boston: Walter de Gruyter).

Orlandi, Giovanni (1985), "La tradizione del 'Physiologus' e i prodromi del bestiario latino," *L'uomo di fronte al mondo animale nell'alto medioevo, vol. 2* (Spoleto: Presso la Sede del Centro), 1057–1106.

Osborne, Catherine (2007), *Dumb Beasts and Dead Philosophers. Humanity and the Humane in Ancient Philosophy and Literature* (Oxford: Oxford University Press).

Overgaard, Søren (2015), "The Unobservability Thesis," *Synthese* 194/3, 1–18.

Page, Sophie (2007), "Good Creation and Demonic Illusions," *A Cultural History of Animals in the Medieval Age*, ed. Brigitte Resl (Oxford/New York: Berg), 27–57.

Panaccio, Claude (2010), "Mental Representation," *The Cambridge Companion to Medieval Philosophy*, vol. 1, ed. Robert Pasnau (Cambridge/New York: Cambridge University Press), 346–356.

Pascua, Esther (2007), "From Forest to Farm and Town: Domestic Animals from ca. 1000 to ca. 1450," *A Cultural History of Animals in the Medieval Age*, ed. Brigitte Resl (Oxford/New York: Berg), 81–102.

Pasnau, Robert (1997), *Theories of Cognition in the Later Middle Ages* (Cambridge: Cambridge University Press).

Pasnau, Robert (2002), *Thomas Aquinas on Human Nature. A Philosophical Study of* Summa theologiae *Ia 75–89* (Cambridge: Cambridge University Press).

Perälä, Mika (2014), "Medieval Theories of Judgement and Reasoning," *Sourcebook for the History of the Philosophy of Mind. Philosophical Psychology from Plato to Kant*, eds. Simo Knuuttila and Juha Sihvola (Dordrecht: Springer Netherlands), 323–334.

Perfetti, Stefano (2004), "I libri *De animalibus* di Aristotele e i saperi sugli animali nel XIII secolo," *Parva naturalia. Saperi medievali, natura e vita*, eds. Chiara Crisciani, Roberto Lambertini, and Romana Martorelli (Pisa: Istituti editoriali e poligrafici internazionali), 143–170.

Perler, Dominik (1990), *Satztheorien. Texte zur Sprachphilosophie und Wissenschaftstheorie im 14. Jahrhundert* (Darmstadt: Wissenschaftliche Buchgesellschaft).

Perler, Dominik (1992). *Der propositionale Wahrheitsbegriff im 14. Jahrhundert* (Berlin: Walter de Gruyter).

Perler, Dominik (2004), *Theorien der Intentionalität im Mittelalter*, 2nd ed. (Frankfurt am Main: Klostermann).

Perler, Dominik (2006), "Intentionality and Action. Medieval Discussions on the Cognitive Capacities of Animals," *Intellect et imagination dans la Philosophie Médiévale*, vol. 1, eds. Maria C. Pacheco and José F. Meirinhos (Turnhout: Brepols), 72–98.

Perler, Dominik (2008), "Seeing and Judging: Ockham and Wodeham on Sensory Cognition," *Theories of Perception in Medieval and Early Modern Philosophy*, eds. Simo Knuuttila and Pekka Kärkkäinen (Dordrecht: Springer), 151–169.

Perler, Dominik (2009), "Graduelle oder kategorische Unterschiede? Leibniz über das Verhältnis von Tieren und Menschen," *Leibniz neu denken*, eds. Erich Barke, Rolf Wernstedt, and Herbert Breger (Stuttgart: Steiner), 75–95.

Perler, Dominik (2011), *Transformationen der Gefühle. Philosophische Emotionstheorien 1270–1670* (Frankfurt am Main: Fischer).

Perler, Dominik (2012a), "Die kognitive Struktur von Hoffnung. Zwei mittelalterliche Erklärungsmodelle," *Deutsche Zeitschrift für Philosophie* 60/1, 73–89.

Perler, Dominik (2012b), "Why is the Sheep Afraid of the Wolf? Medieval Debates on Animal Passions," *Emotion and Cognitive Life in Medieval and Early Modern Philosophy*, eds. Martin Pickavé and Lisa Shapiro (Oxford: Oxford University Press), 32–52.

Perler, Dominik (2014a), "Can We Trust Our Senses? Fourteenth-Century Debates on Sensory Illusions," *Uncertain Knowledge. Scepticism, Relativism, and Doubt in the Middle Ages*, eds. Dallas G. Denery II, Kantik Ghosh, and Nicolette Zeeman (Turnhout: Brepols), 63–90.

Perler, Dominik (2014b), "Spinoza über Tiere," *Archiv für Geschichte der Philosophie* 96/2, 232–261.

Perler, Dominik (ed.) (2015a), *The Faculties. A History* (Oxford: Oxford University Press).

Perler, Dominik (2015b), "Faculties in Medieval Philosophy," *The Faculties. A History*, ed. id. (Oxford: Oxford University Press), 97–139.

Perler, Dominik (2015c), "Perception in Medieval Philosophy," *The Oxford Handbook of the Philosophy of Perception*, ed. Mohan Matthen (Oxford: Oxford University Press), 51–65.

Perler, Dominik, and Markus Wild (eds.) (2005), *Der Geist der Tiere. Philosophische Texte zu einer aktuellen Diskussion* (Frankfurt am Main: Suhrkamp).

Petrus, Klaus, and Markus Wild (eds.) (2013), *Animal Minds & Animal Ethics. Connecting Two Separate Fields* (Berlin/Bielefeld: Transcript).

Piro, Francesco (2005), "Animalité et connaissance animale. Notes sur la doctrine de l'estimative et sur ses interprétations en Occident," *Animal et animalité dans la philosophie de la Renaissance et de l'Âge Classique*, ed. Thierry Gontier (Leuven: Peeters), 131–151.

Pluta, Olaf (1996), "Der Alexandrismus an den Universitäten im späten Mittelalter," *Bochumer Philosophisches Jahrbuch für Antike und Mittelalter* 1/1, 81–109.

Pluta, Olaf (2002a), "John Buridan on Universal Knowledge," *Bochumer Philosophisches Jahrbuch für Antike und Mittelalter* 7, 25–46.

Pluta, Olaf (2015), "Mental Representation in Animals and Humans. Some Late Medieval Discussions," *Intentionality, Cognition, and Mental Representation in Medieval Philosophy*, ed. Gyula Klima (New York: Fordham University Press), 273–286.

Pormann, Peter E. (2013), "Avicenna on Medical Practice, Epistemology, and the Physiology of the Inner Senses," *Interpreting Avicenna. Critical Essays*, ed. Peter Adamson (Cambridge: Cambridge University Press), 91–108.

Povinelli, Daniel J., and Steve Giambrone (1999), "Inferring Other Minds: Failure of the Argument from Analogy," *Philosophical Topics* 27/1, 167–201.

Preece, Rod (2002), *Awe for the Tiger, Love for the Lamb. A Chronicle of Sensibility to Animals* (New York/London: Routledge).

Premack, David, and Guy Woodruff (1978), "Does the Chimpanzee Have A Theory of Mind?," *Behavioral and Brain Sciences* 1/4, 515–526.

Reina, Maria E. (1986), "Un abozzo di polemica sulla psicologia animale: Gregorio da Rimini contro Adamo Wodeham," *L'homme et son univers au Moyen Âge*, ed. Christian Wenin (Leuven: Publications de l'Institut Supérieur de Philosophie), 598–609.

Rescorla, Michael (2009), "Chrysippus' Dog as a Case Study in Non-Linguistic Cognition," *The Philosophy of Animal Minds*, ed. Robert W. Lurz (Cambridge: Cambridge University Press), 52–71.

Resl, Brigitte (2007a), "Animals in Culture, ca. 1000–ca. 1400," *A Cultural History of Animals in the Medieval Age*, ed. id. (Oxford/New York: Berg), 1–26.

Resl, Brigitte (2007b), "Beyond the Ark: Animals in Medieval Art," *A Cultural History of Animals in the Medieval Age*, ed. id. (Oxford/New York: Berg), 179–201.

Resnick, Irven M., and Kenneth F. Kitchell Jr. (1996), "Albert the Great on the 'Language' of Animals," *American Catholic Philosophical Quarterly* 70/1, 41–61.

Resnick, Irven M., and Kenneth F. Kitchell Jr. (1999), "Introduction: The Life and Works of Albert the Great," *Albertus Magnus On Animals. A Medieval Summa Zoologica*, vol. 1, ed. and tr. id. (Baltimore: The Johns Hopkins University Press), 1–42.

Rheins, Jason G. (2015), "*Homo numerans, venerans,* or *imitans*? Human and Animal Cognition in *Problemata* 30.6," *The Aristotelian* Problemata Physica. *Philosophical and Scientific Investigations*, ed. Robert Mayhew (Leiden: Brill), 381–412.

Rignani, Orsola (2006), "Internal and External Senses in Roger Bacon," *Intellect et imagination dans la Philosophie Médiévale*, vol. 2, eds. Maria C. Pacheco and José F. Meirinhos (Turnhout: Brepols), 1219–1229.

Roberts, Robert C. (1992), "Thomas Aquinas on the Morality of Emotions," *History of Philosophy Quarterly* 9/3, 287–305.

Roling, Bernd (2008), *Locutio angelica. Die Diskussion der Engelsprache als Antizipation einer Sprechakttheorie in Mittelalter und Früher Neuzeit* (Leiden/Boston: Brill).

Roling, Bernd (2010), *Drachen und Sirenen. Die Rationalisierung und Abwicklung der Mythologie an den europäischen Universitäten* (Leiden/Boston: Brill).

Roling, Bernd (2011), "*Syllogismus brutorum*: Die Diskussion der animalischen Rationalität bei Albertus Magnus und ihre Rezeption im Mittelalter und in der Frühen Neuzeit," *Recherches de Théologie et Philosophie médiévales* 78/1, 221–275.

Roling, Bernd (2013a), "Animalische Sprache und Intelligenz im Schriftkommentar. Bileams Esel in der Bibelkommentierung des Mittelalters und der Neuzeit," *Neo-Latin Commentaries and the Management of Knowledge in the Late Middle Ages and the Early Modern Period (1400–1700)*, eds. Karl A.E. Enenkel and Henk Nellen (Leuven: Leuven University Press), 409–443.

Roling, Bernd (2013b), "Die Geometrie der Bienenwabe: Albertus Magnus, Karl von Baer und die Debatte über das Vorstellungsvermögen und die Seele der Insekten zwischen Mittelalter und Neuzeit," *Recherches de Théologie et Philosophie médiévales* 80/2, 401–504.

Roling, Bernd (2014), "Die Diskussion der *anima brutorum* an den protestantischen Universitäten und ihre Hintergründe," *Wolfenbütteler Renaissance-Mitteilungen* 34/2, 213–230.

Roling, Bernd (2015), "Die Gleichzeitigkeit aller Gedanken. Debatten zur Erkenntnis der ‚anima separata' zwischen Mittelalter und Früher Neuzeit," *Gleichzeitigkeit. Narrative Synchronisierungsmodelle in der Literatur des Mittelalters und der Frühen Neuzeit*, eds. Susanne Köbele and Coralie Rippl (Würzburg: Königshausen & Neumann), 53–74.

Rosier-Catach, Irène (2006), "'Solo all'uomo fu dato di parlare'. Dante, gli angeli e gli animali," *Rivista di filosofia neo-scolastica* 3/98, 435–465.

Rubini, Paolo (2015), *Pietro Pomponazzis Erkenntnistheorie. Naturalisierung des menschlichen Geistes im Spätaristotelismus* (Leiden: Brill).

Sabra, Abdelhamid I. (1978), "Sensation and Inference in Alhacen's Theory of Visual Perception," *Studies in Perception. Interrelations in the History of Philosophy and Science*, eds. Peter K. Machamer and Robert G. Turnbull (Columbus: Ohio State University Press), 160–185.

Salisbury, Joyce E. (2011), *The Beast Within. Animals in the Middle Ages*, 2nd ed. (New York/London: Routledge).

Sander, Christoph (2016), "Der Dämon im Text: Lateinische Lesarten von *De somno* 453b22 und *De divinatione per somnum* 463b12 zwischen 1150 und 1650," *Recherches de Théologie et Philosophie médiévales*, 83/2, 245–311.

Schäfer, Christian, and Martin Thurner (eds.) (2013), Passiones animae. *Die 'Leidenschaften der Seele' in der mittelalterlichen Theologie und Philosophie*, 2nd ed. (Berlin: Akademie Verlag).

Schmidtke, Dietrich (1968), *Geistliche Tierinterpretation in der deutschsprachigen Literatur des Mittelalters (1100–1500)*, 2 vols. (Berlin: Ernst-Reuter-Gesellschaft).

Schmidtke, Dietrich (1993), "Geistliche Tierinterpretation," *Geistliche Aspekte mittelalterlicher Naturlehre*, ed. Benedikt K. Vollmann (Wiesbaden: Dr. Ludwig Reichert), 26–39.

Schoeler, Gregor (2005), "Poetischer Syllogismus – bildliche Redeweise – Religion. Vom aristotelischen *Organon* zu al-Fārābīs Religionstheorie," *Logik und Theologie. Das Organon im arabischen und im lateinischen Mittelalter*, eds. Dominik Perler and Ulrich Rudolph (Leiden/Boston: Brill), 45–58.

Schoeler, Gregor (2013), "The 'Poetic Syllogism' Revisited," *Oriens* 41/1–2, 1–26.

Searle, John (2002), "Can Computers Think?," *Philosophy of Mind. Classical and Contemporary Readings*, ed. David Chalmers (Oxford: Oxford University Press), 659–675.

Singer, Peter (2006), "Morality, Reason, and the Rights of Animals," *Primates and Philosophers. How Morality Evolved*, eds. Stephen Macedo and Josiah Ober (Princeton/Oxford: Princeton University Press), 140–158.

Smets, An, and Baudouin van den Abeele (2007), "Medieval Hunting," *A Cultural History of Animals in the Medieval Age*, ed. Brigitte Resl (Oxford/New York: Berg), 59–79.

Smith, A. Mark (2001), "Introduction," *Alhacen's Theory of Visual Perception. A Critical Edition, with English Translation and Commentary, of the First Three Books of Alhacen's* De aspectibus, *the Medieval Latin Version of Ibn al-Haytham's Kitāb al-Manāẓir*, vol. 1, ed. id. (Philadelphia: American Philosophical Society), xv–cliv.

Smith, A. Mark (2010), "Perception," *The Cambridge Companion to Medieval Philosophy*, vol. 1, ed. Robert Pasnau (Cambridge/New York: Cambridge University Press), 335–345.

Sober, Elliott (2000), "Evolution and the Problem of Other Minds, *The Journal of Philosophy* 97/7, 365–386.
Sober, Elliott (2005), "Comparative Psychology Meets Evolutionary Biology. Morgan's Canon and Cladistic Parsimony," *Thinking with Animals. New Perspectives on Anthropomorphism*, eds. Gregg Mitman and Lorraine Daston (New York: Columbia University Press), 85–99.
Sobol, Peter G. (1993), "The Shadow of Reason: Explanations of Intelligent Animal Behaviour in the Thirteenth Century," *The Medieval World of Nature. A Book of Essays*, ed. Joyce E. Salisbury (New York/London: Garland), 109–128.
Sorabji, Richard (1993a), "Animal Minds," *Southern Journal of Philosophy* 31, 1–18.
Sorabji, Richard (1993b), *Animal Minds and Human Morals. The Origins of the Western Debate* (London: Duckworth).
Speer, Andreas (1995), *Die entdeckte Natur. Untersuchungen zu Begründungsversuchen einer 'scientia naturalis' im 12. Jahrhundert* (Leiden/New York/Köln: Brill).
Speer, Andreas (2001), "Vernunft; Verstand (III. Mittelalter – D. Hochscholastik)," *Historisches Wörterbuch der Philosophie*, vol. 11 (Darmstadt: Wissenschaftliche Buchgesellschaft), 780–786.
Stach, Silke (2004), "Pattern Recognition in Honeybees. Generalization and Categorization," PhD. Diss. (Berlin: Freie Universität).
Starzak, Tobias (2014), *Kognition bei Menschen und Tieren. Eine vergleichende philosophische Perspektive* (Berlin/München/Boston: Walter de Gruyter).
Steel, Carlos G., Guy Guldentops, and Pieter Beullens (eds.) (1999), *Aristotle's Animals in the Middle Ages and Renaissance* (Leuven: Leuven University Press).
Steiner, Gary (2005), *Anthropocentrism and Its Discontents. The Moral Status of Animals in the History of Western Philosophy* (Pittsburgh: Pittsburgh University Press).
Steiner, Gary (2008), *Animals and the Moral Community. Mental Life, Moral Status, and Kinship* (New York: Columbia University Press).
Steneck, Nicholas H. (1974), "Albert the Great on the Classification and Localization of the Internal Senses," *Isis* 65/2, 193–211.
Stump, Eleonore (1999), "Aquinas on the Mechanisms of Cognition: Sense and Phantasia," *Medieval Analyses of Language and Cognition*, eds. Sten Ebbesen and Russell L. Friedman (Copenhagen: C.A. Reitzels), 377–395.
Stump, Eleonore (2003), *Aquinas* (London/New York: Routledge).
Sudhoff, Walther (1913), "Die Lehre von den Hirnventrikeln in textlicher und graphischer Tradition des Altertums und Mittelalters," *Archiv für Geschichte der Medizin* 7/3, 149–205.
Summers, David (1987), *The Judgment of Sense. Renaissance Naturalism and the Rise of Aesthetics* (Cambridge/New York/Melbourne: Cambridge University Press).
Tachau, Katherine H. (1988), *Vision and Certitude in the Age of Ockham. Optics, Epistemology and the Foundations of Semantics 1250–1345* (Leiden et al.: Brill).

Tachau, Katherine H. (1993), "What Senses and Intellect Do: Argument and Judgment in Late Medieval Theories of Knowledge," *Argumentationstheorie. Scholastische Forschungen zu den logischen und semantischen Regeln korrekten Folgerns*, ed. Klaus Jacobi (Leiden: Brill), 653–668.

Tellkamp, Jörg A. (1999), *Sinne, Gegenstände und Sensibilia. Zur Wahrnehmungslehre des Thomas von Aquin* (Leiden/Boston/Köln: Brill).

Tellkamp, Jörg A. (2006), "The *sensibilia per accidens* according to Thomas Aquinas," *Intellect et imagination dans la Philosophie Médiévale*, vol. 3, eds. Maria C. Pacheco and José F. Meirinhos (Turnhout: Brepols), 1351–1361.

Tellkamp, Jörg A. (2012a), "Albert the Great on Structure and Function of the Inner Senses," *The Judeo-Christian-Islamic Heritage. Philosophical and Theological Perspectives*, eds. Richard C. Taylor and Irfan A. Omar (Milwaukee: Marquette University Press), 305–324.

Tellkamp, Jörg A. (2012b), "*Vis aestimativa* and *vis cogitativa* in Thomas Aquinas's *Commentary on the Sentences*," *The Thomist* 76, 611–640.

Tellkamp, Jörg A. (2013), "Albert the Great on Perception and Non-Conceptual Content," *Philosophical Psychology in Arabic Thought and the Latin Aristotelianism of the 13th Century*, eds. Luis Xavier López-Farjeat and Jörg Alejandro Tellkamp (Paris: Vrin), 205–221.

Tellkamp, Jörg A. (2016), "Aping Logic? Albert the Great on Animal Mind and Action," *Subjectivity and Selfhood in Medieval and Early Modern Philosophy*, eds. Jari Kaukua and Tomas Ekenberg (Dordrecht: Springer), 109–123.

Thijssen, Johannes M.M.H. (1995), "Reforging the Great Chain of Being. The Medieval Discussion of the Human Status of 'Pygmies' and Its Influence on Edward Tyson," *Ape, Man, Apeman. Changing Views Since 1600*, eds. Raymond Corbey and Bert Theunissen (Leiden: Department of Prehistory, Leiden University), 43–59.

Toivanen, Juhana (2007), "Peter Olivi on Internal Senses," *British Journal for the History of Philosophy* 15/3, 427–454.

Toivanen, Juhana (2011), "Peter of John Olivi on the Psychology of Animal Action," *Journal of the History of Philosophy* 49/4, 413–438.

Toivanen, Juhana (2013a), *Perception and the Internal Senses. Peter of John Olivi on the Cognitive Functions of the Sensitive Soul* (Leiden: Brill).

Toivanen, Juhana (2013b), "Perceptual Self-Awareness in Seneca, Augustine, and Olivi," *Journal of the History of Philosophy* 51(3), 355–382.

Toivanen, Juhana (2016), "Beasts, Human Beings, or Gods: Human Subjectivity in Medieval Political Philosophy," *Subjectivity and Selfhood in Medieval and Early Modern Philosophy*, eds. Jari Kaukua and Tomas Ekenberg (Dordrecht: Springer), 181–197.

Toivanen, Juhana (forthcoming), "Marking the Boundaries: Animals in Medieval Latin Philosophy," *Animals. A History*, eds. G. Fay Edwards and Peter Adamson (Oxford: Oxford University Press).

Tomasello, Michael (2014), *A Natural History of Human Thinking* (Cambridge, Mass./London: Harvard University Press).

Tomasello, Michael, and Josep Call (1997), *Primate Cognition* (New York/Oxford: Oxford University Press).

Van Dusen, David (2014), *The Space of Time. A Sensualist Interpretation of Time in Augustine, Confessions X to XII* (Leiden/Boston: Brill).

Voisenet, Jacques (2000), *Bêtes et hommes dans le monde médiéval. Le bestiaire des clercs du Ve au XIIe siècle* (Turnhout: Brepols).

Wallach, Wendell, and Colin Allen (2008), *Moral Machines. Teaching Robots Right from Wrong* (Oxford/New York: Oxford University Press).

Wannenmacher, Julia E. (2017), "Von koranischer Überlieferung und zeitgemäßen Perspektiven: Tiere und Tierethik im Islam (I)," *Freiburger Zeitschrift für Philosophie und Theologie* 64/1, 170–194.

Wasserman, Edward A., and Thomas R. Zentall (2006), "Comparative Cognition: A Natural Science Approach to the Study of Animal Intelligence," *Comparative Cognition. Experimental Explorations of Animal Intelligence*, eds. id. (Oxford/New York: Oxford University Press), 3–11.

Wild, Markus (2006), *Die anthropologische Differenz. Der Geist der Tiere in der frühen Neuzeit bei Montaigne, Descartes und Hume* (Berlin: Walter de Gruyter).

Wild, Markus (2013), *Tierphilosophie zur Einführung*, 3rd corr. ed. (Hamburg: Junius).

White, Terence H. (2002), *The Book of Beasts. Being a Translation from a Latin Bestiary of the Twelfth Century* (Madison: Parallel Press).

Wynne, Clive D.L. (2007), "What are Animals? Why Anthropomorphism is Still Not a Scientific Approach to Behavior," *Comparative Cognition & Behavior Reviews* 2, 125–135.

Wolfson, Harry Austryn (1935), "The Internal Senses in Latin, Arabic, and Hebrew Philosophic Texts," *The Harvard Theological Review* 28/2, 69–133.

Wood, Rega (2007), "Imagination and Experience in the Sensory Soul and Beyond: Richard Rufus, Roger Bacon and Their Contemporaries," *Forming the Mind. Essays on the Internal Senses and the Mind/Body Problem from Avicenna to the Medical Enlightenment*, ed. Henrik Lagerlund (Dordrecht: Springer), 27–57.

Yamamoto, Dorothy (1998), "Aquinas and Animals: Patrolling the Boundary?," *Animals on the Agenda. Questions about Animals for Theology and Ethics*, eds. Andrew Linzey and Dorothy Yamamoto (London: SCM Press), 80–89.

Yates, Francis A. (1969), *The Art of Memory* (Harmondsworth: Penguin Books).

Zupko, Jack (2008), "Horse Sense and Human Sense: The Heterogeneity of Sense Perception in Buridan's Philosophical Psychology," *Theories of Perception in Medieval and Early Modern Philosophy*, eds. Simo Knuuttila and Pekka Kärkkäinen (Dordrecht: Springer), 171–186.

Index of Names

Index of Subjects

www.ingramcontent.com/pod-product-compliance
Lightning Source LLC
Chambersburg PA
CBHW071253180926
51613CB00001B/27
9789004363625